LINING UP DATA in ArcGIS

a guide to map projections

MARGARET M. MAHER

ESRI Press
REDLANDS, CALIFORNIA

Esri Press, 380 New York Street, Redlands, California 92373-8100

 First edition 2010

14 13 12 11 10   2 3 4 5 6 7 8 9 10

Printed in the United States of America

Library of Congress Cataloging-in-Publication Data
Maher, Margaret M.
Lining up data in ARCGIS : a guide to map projections / Margaret M. Maher.—1st ed.
p.   cm.
Includes bibliographical references.
ISBN 978-1-58948-249-4 (pbk. : alk. paper)
1. ArcGIS. 2. Geographic information systems. 3. Graphical user interfaces (Computer systems) I. Title.
G70.212.M275 2010
910.285—dc22   2010000193

Ask for Esri Press titles at your local bookstore or order by calling 800-447-9778, or shop online at www.esri.com/esripress.
Outside the United States, contact your local Esri distributor or shop online at www.eurospanbookstore.com/Esri.
Esri Press titles are distributed to the trade by the following:

In North America:
Ingram Publisher Services
Toll-free telephone: 800-648-3104
Toll-free fax: 800-838-1149
E-mail: customerservice@ingrampublisherservices.com

In the United Kingdom, Europe, Middle East and Africa, Asia, and Australia:
Eurospan Group
3 Henrietta Street
London WC2E 8LU
United Kingdom
Telephone: 44(0) 1767 604972
Fax: 44(0) 1767 601640
E-mail: eurospan@turpin-distribution.com

CONTENTS

PREFACE

This practical guide will help you line up your data in a map projection as easily as possible. It is a working manual intended to lead you quickly to the solution when data alignment problems arise. Many scholarly publications delve deeply into the mathematics of map projections, detailing the calculations used behind the scenes to project data from the curved surface of the earth onto a flat piece of paper. Because we reside on a lumpy ball of minerals suspended in space, even a simple concept as the "center" of the earth is the subject of much debate and reams of complex calculations. A search on the Internet will turn up hundreds of thousands of Web pages devoted to such scholarly work.

This book is not intended for the geodesist, highly trained in mathematics, who has a profound understanding of map projections. It is for those working in the field of geographic information systems (GIS) and others in the process of making a map who, like so many who have called me over the years for technical support, sometimes run into a problem lining up their data in ArcMap.

ArcMap is the application for making maps and analyzing data within ArcGIS Desktop, which is made by ESRI and is the leading GIS software. While this guide is written with the users of ArcGIS Desktop specifically in mind, the basic principles of map projection apply and will be useful to others. I have worked in Support Services at ESRI for ten years and have had the pleasure of talking with thousands of you while closing more than 12,000 support incidents. The majority of these telephone calls have been questions about lining up data in ArcMap.

Many of you have requested a book that would provide comprehensive information about working with coordinate systems and aligning data. So here it is, the book that will answer most of your day-to-day questions and demystify some of the complex issues surrounding map projections. It is organized in an especially practical way, on the assumption that you may be in the midst of a dilemma right now and very much in need of a way to resolve it quickly.

My thanks go especially to Melita Kennedy, David Burrows, and Rob Juergens, whose endless patience and understanding have enabled me to distill the contents from the information they have so generously shared. I am also very grateful to Philip Sanchez, Jeffrey Reinhart, and Alex LeReaux who have taught me about computer-aided design (CAD) software and how to work with data created in some of these applications.

I am most grateful to the thousands of ESRI software users who inspired this book. Working with ArcGIS users in Support Services is very rewarding for me. This book is a way for me to share with you some of the enjoyment and excitement in learning and communication that has taken place over the past ten years. I wish you the best in all your endeavors.

ABOUT THE AUTHOR

Margaret M. Maher specializes in map projections and data conversion in Support Services at ESRI, maker of the leading geographical information systems (GIS) software. She holds a bachelor of science degree in a specialized major combining studies in GIS and geology from California State University, Sacramento. Maher has written numerous articles for the ESRI Support Services Knowledge Base (of the twelve most-accessed articles pertaining to the operation of ArcGIS Desktop, she wrote ten). Maher lives in Cherry Valley, California, with her daughter, Julie, and husband Georges Dib.

I dedicate this book to my daughter, Julie,
who makes *my* world line up.

INTRODUCTION

"My boss told me to make a map using ArcMap, but when I add the data, I get an error message that says 'missing spatial reference' and the data doesn't line up. I've got a deadline! How do I fix it?" (Chapter 1)

"When I add my new data to ArcMap, I get a Warning box that says 'Inconsistent extent' and the data doesn't line up. What is the problem?" (Chapter 1)

"How can I tell what datum my data is on? I tried to find out from the data source, but they didn't know." (Chapter 2)

"I received a parcel shapefile from the county, but it doesn't line up with my other data in ArcMap. What now?" (Chapter 3)

"The CAD file my client sent me is huge when I add it to ArcMap with my other data. Why is that?" (Chapter 4)

"A CAD file I received from a client draws way off from my other data in ArcMap. What should I do?" (Chapter 5)

"When I add a client's CAD file to ArcMap it draws at an angle. How can I fix that?" (Chapter 6)

"I have some new data, but when I add it to ArcMap with my other data, a message box pops up that says 'Geographic Transformation Warning' and the data is off by about 200 feet. What am I supposed to do?" (Chapter 7)

"When I add data to ArcMap, and try to set the datum transformation, I get eight different options. Which one am I supposed to pick?" (Chapter 8)

"What projection should I use for my data?" (Chapter 9)

"What do all these items in a projection file mean, and what do they do to the data?" (Chapter 10)

"When I add x,y data to ArcMap, the points draw in the wrong place. Why?" (Chapter 10)

"Why aren't my buffers round in ArcMap?" (Chapter 10)

Sound familiar?

This best practices book addresses the questions and problems above, which are the ones heard most often over the last decade in phone calls to ESRI Support Services. The book also covers the basics about working with coordinate systems and map projections. While this guide is written for users of ArcGIS Desktop working with vector data, the basic principles also apply to raster image formats. Because there are substantial differences between the methods for handling vector and raster data formats in ArcGIS Desktop, we had to make a choice: screenshots and descriptions of procedures will apply only to vector data.

Many data alignment problems occur when new data is received from sources outside your organization or department. The **cardinal rule** in resolving these data alignment issues is to **keep the data in the original format until the spatial reference has been identified and the projection defined correctly**. Importing shapefiles, CAD data, or x,y data into a geodatabase before identifying the spatial reference of the data introduces additional complexity into the process. As long as the data can be displayed in ArcMap in the original format, you can avoid complications by keeping your new data in the original format until a correct coordinate system definition can be applied to the data.

For the most part, examples of coordinate systems and screenshots of sample data are taken from the United States. Many countries like Australia, Canada, Colombia, France, Germany, Japan, and others also use different coordinate systems that are specifically designed to minimize distortion within national borders, or coordinate systems that are applied to specific regions within the country. The instructions and procedures described in this book can be applied to align data for any country in the world, but a comprehensive treatment of data alignment issues worldwide is beyond the scope of what can be accomplished here, so we are using U.S. examples.

This book is meant to be a practical working manual. Reading it cover to cover will give you lots of useful information to answer the questions posed here and to resolve additional related issues. In your first reading, you might like to mark the pages that explain how to fix situations you often encounter. You also can use this book to rapidly find solutions when problems arise. For the most part, the book is arranged so that each chapter addresses a specific question or problem and provides solutions for that issue. You can refer to the table of contents and the index to locate the information you need.

Chapters 1, 2, and 3 contain steps to identify the coordinate system of data in ArcMap, and define the coordinate system to match the data, if the data has been created using a standard projection.

Chapter 4 describes techniques to identify nonstandard units of measure that are sometimes used to create data, and techniques to modify an existing projection file to incorporate these different units.

Chapters 5 and 6 offer instructions for modifying standard projection file parameters, and for creating custom projection files to align data, including CAD files, that was created using local coordinate systems.

Chapters 7 and 8 discuss geographic (datum) transformations in greater detail, explain their purpose, and describe the various transformation methods that are supported in ArcGIS Desktop. Many scholarly sources delve into the mathematics of these transformation methods. You won't find that intricacy here. However, you will find what you need to know to help your data line up: the names of the supported transformation methods and the parameters or files required to apply each method in ArcGIS Desktop.

Chapter 9 discusses the different properties of various map projections, their uses and applications, and methods for determining which map projection will be appropriate for a specific project and for storing data.

Chapter 10 discusses the parameters included in a projection file, how values for those parameters are determined, and what purpose each parameter serves. This chapter also includes detailed information about adding x,y data to ArcMap, identifying the coordinate system of the data in the table, and then converting that data to a shapefile or geodatabase feature class. This chapter also addresses the frequently asked question about the shape of buffers displayed in the ArcMap data frame.

You will notice as you read that some information is repeated in different chapters. Each chapter is part of the whole, but also stands alone, addressing specific issues related to working with coordinate systems. This format allows you to locate the solution for a specific problem without referring to instructions or examples in other chapters. Restating important information in various chapters will also allow you the opportunity to review concepts presented in various ways and from different perspectives, and integrate these concepts into a better understanding of coordinate systems.

To facilitate the flow of your reading, refer to appendix A at the back of the book. This appendix contains references to a number of Knowledge Base articles from the ESRI Support Center. Some of the articles contain download links referred to in this text. Downloading copies of the relevant materials from ESRI Knowledge Base articles 29280, 21327, and 24646 before you begin will ensure that you have the lists and tables at hand when you find references to them in the text.

Appendix B contains references to default installation locations for ArcGIS Desktop based on the software version and Windows operating system on your computer. Appendix C contains default paths to user profiles, again depending on the ArcGIS Desktop version and Windows operating system. You will need this information as you read and work in order to locate projection files (PRJ) stored on your computer and to copy projection files from one directory to another.

Note for those using the new version of the software: don't worry when you see that the screenshots included in the text are from ArcGIS Desktop version 9.3. Though changes have been made to toolbars and icons in the latest version of ArcGIS Desktop, the *general* appearance of dialog boxes and functionality of tools remains similar enough for you to follow along just fine.

There is no getting around the fact that understanding coordinate systems is a challenge. But it is one you can meet. This book grew out of interactions with thousands of GIS users over a decade, many of whom asked me to summarize in an e-mail what we had just discussed over the phone. It was in honing those instructions over the years—finding effective ways to phrase them that made the most sense—that this content evolved into the practical guide you have here, one you can rely on 24/7.

CHAPTER 1

IDENTIFYING THE TYPE OF COORDINATE SYSTEM FOR DATA USING ARCMAP

"My boss told me to make a map using ArcMap, but when I add the data, I get an error message that says 'missing spatial reference' and the data doesn't line up. I've got a deadline! How do I fix it?"

"When I add my new data to ArcMap, I get a Warning box that says 'Inconsistent extent' and the data doesn't line up. What is the problem?"

The coordinate system for data provides a frame of reference so that users of geographic information systems (GIS) can identify the location of features on the surface of the earth, align data, and create maps. These maps enable users to perform spatial analyses of the data and view its relationship to other features.

Dozens of map projections, the means for displaying features from the curved surface of the earth to a flat sheet of paper, have been calculated by geodesists — scientists who study the shape of the earth. Each of these map projections has been calculated to preserve one or more specific properties of the data — shape, distance, area, or direction. Literally, an infinite number of specific projections can be created for data depending on the extent, location, and particular property of the data that is most important for a specific project, general data storage, and maintenance.

No wonder people are confused by projections.

This book is organized to help you to narrow the huge number of possible coordinate systems for data to a manageable selection and to identify the unknown coordinate systems by following the methods outlined. The book also provides instructions for creating custom projection files, if data is not in a standard coordinate system, so that data will properly line up in ArcMap.

All data is created in some coordinate system. Once you identify which coordinate system your data is in, you can correctly select (define) the map projection file that will render your data in the right location in ArcMap in relation to other data. The first three chapters contain steps to identify the coordinate system of data in ArcMap, and define the coordinate system to match the data, if the data has been created using a standard projection.

FEATURES IN VECTOR DATASETS

Vector datasets contain point, line, or polygon features that illustrate the position of real or imaginary features on the earth. Water or oil wells are real features that would be maintained in the vector dataset as points. A road is a real feature: the centerline can be maintained as a line feature in the vector dataset. A tax parcel is an imaginary polygon feature, with boundaries and an area defined by national, state, or local laws. Attribute tables for the data store relevant information about the features in the dataset.

Vast quantities of free vector data are available for download from the Internet. Data exchange between GIS analysts is a daily occurrence. Data collected with a Global Positioning System (GPS) unit is part of many users' workflows. Metadata describes the source of the data, collection methods, relative accuracy of the data, and coordinate system or projection of the data, but in many cases metadata is not provided.

In order to use vector data in ArcMap, the coordinate system for the data must be *identified* based on the coordinate extent of the data as displayed in ArcMap > Layer Properties > Source tab. Since ArcGIS Desktop installs more than 4,000 projection files with the software, and only *one* of those projection files properly describes the coordinate system of a specific dataset, the coordinate system of the dataset *must* be correctly identified so that the proper projection file (PRJ) is applied to the data. The projection must then be correctly *defined* by applying the projection that describes the coordinate system of the data. ArcMap's project-on-the-fly utility, used throughout the book, helps identify projections to align data.

The difference between defining a coordinate system and projecting data

All GIS data is created in some coordinate system. All GIS data covers some extent on the surface of the earth, whether the data is points, lines, polygons, annotation, or some other type of feature. The extent coordinates can be in decimal degrees, feet, meters, tenths of an inch, tenths of a foot, millimeters, centimeters, or kilometers—the list is endless. When the data is added to ArcMap, and you right-click the name of the data layer, select Properties > Source tab, and look at the numbers in the Extent box at the top of the tab, the Top, Bottom, Left, and Right coordinates are the extent of the data, in the units and coordinate system used to create the data.

The coordinate system of the data is defined in relation to the units in the Extent box. This book is about identifying the coordinate system for data, so that the coordinate system can be correctly defined and the data will draw in the right location in ArcMap in relation to other data.

ArcGIS Desktop is installed with more than 4,000 projection files. In order to define the coordinate system of the data correctly, you must first identify the projection file that correctly describes the coordinates for the data. You cannot randomly select one of these 4,000 projection files to define the data. You have to pick the one that correctly describes the coordinate extent in the proper units and coordinate system. If none of these projection files matches the properties of the data, you will have to create a custom projection file, as described in chapters 4, 5, and 6.

To define the coordinate system for the data, you will use the Define Projection tool located in ArcToolbox > Data Management Tools > Projections and Transformations. (In ArcGIS Desktop 10, from within ArcMap, select Catalog window > Toolboxes > System Toolboxes to access this path.)

Say you receive data that is in a geographic coordinate system, GCS_North_American_1927. There is no metadata and no projection file, but you use the steps in this book to identify and define the coordinate system as North American Datum 1927.

All your other data is in NAD 1983 StatePlane California VI FIPS 0406 Feet. *You cannot perform any analysis using the data unless it matches your other data's coordinate system.* You cannot define the data as NAD 1983 StatePlane Feet, because the coordinates of the data are in decimal degrees, not feet.

(continued)

The difference between defining a coordinate system and projecting data (continued)

After defining the projection with the coordinate system that matches the data, you will use the Project tool, located in ArcToolbox > Data Management Tools > Projections and Transformations > Feature. The Project tool creates a new copy of the data in the coordinate system you have selected. When the Project tool is finished, you will still have the original input data, in the original coordinate system. The new data will be in the new coordinate system you selected. Add both sets of data to ArcMap, and compare the extents to see how the coordinate system selected for the output changed the extent of the data.

USING PROJECT ON THE FLY IN ARCMAP

The ArcMap data frame adopts the coordinate system definition of the first layer added to a new empty map. If the first data added has the projection correctly defined, other data that has correct coordinate system definitions will be projected on the fly to the coordinate system of the data frame when added to the map document. The newly added data will be displayed in the data frame's current coordinate system.

The project-on-the-fly utility in ArcMap is intended to facilitate mapmaking and cartographic development, but should not be used when analysis is performed with the data or if the data will be edited. Project on the fly makes no permanent change to the data displayed in the map, and is just as accurate as projecting data with the Project tool in ArcToolbox.

If the newly added data does not have a coordinate system defined or if the coordinate system is defined incorrectly, project on the fly cannot work, data will not align, and the mapping project cannot be completed.

IDENTIFYING THE TYPE OF COORDINATE SYSTEM FOR DATA USING ARCMAP

Data can be created in one of three types of coordinate systems:

1. Geographic
2. Projected
3. Local

Data in each type of coordinate system can be identified by examining the extent of the data, as viewed on the Layer Properties > Source tab in ArcMap.

In order to facilitate work with the new data, copy the data to a folder on the local hard drive where you have write permission. The local folder should not have spaces in the folder name or in the path to the folder. After copying the data to the local hard drive, verify that you have read-write access to the data by checking permissions according to the following instructions.

CHECKING AND CHANGING PERMISSIONS ON NEW FILES OR FOLDERS

- Right-click the Start button and select Explore.
- In Windows Explorer, navigate to the folder where the data has been copied, right-click the folder name, and select Properties > General tab.
- If the Read-only Attributes box is checked, clear the check box, click Apply, select "Apply changes to this folder, subfolders and files," click OK, and OK again.
- If the Read-only Attributes box is empty, or is filled with a green square, open the folder and select the files within the folder.
- Right-click the selected files > Properties > General tab. If the Read-only Attributes box is checked, uncheck it, and click OK.

Now that you have write access to the data, open ArcMap with a new, empty map and add the newly acquired data. If several datasets have been received from the same source and the new datasets line up together when added to ArcMap, you know that the new data all has the same spatial reference. To reduce drawing time, you can turn off all but one of the new data layers while working to identify the coordinate system for the data.

SOME COMMON ERROR MESSAGES AND WARNINGS

If the data does not have a coordinate system defined, the Unknown Spatial Reference message box, shown in figure 1-1, will be displayed when the data is added to ArcMap. Click OK on the dialog box to add the data to the ArcMap window. When working with new data to identify the coordinate system, you want this warning to appear so that various projection files can be tested to find the correct projection definition.

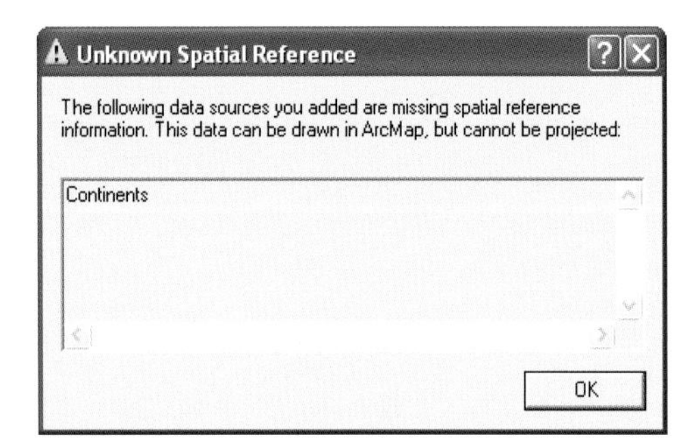

Figure 1-1 Unknown Spatial Reference warning. The data does not have a projection definition. This is the proper condition that should exist while working through these chapters to identify the coordinate system for data.

If instead the "Warning, inconsistent extent!" error message (shown in figure 1-2) is displayed when the data is added, you know that the data does have a defined coordinate system but the coordinate system is wrong for that data. Specifically, the data has been defined with a geographic coordinate system, but the values for the coordinate extent of the data are too large, and the data is not in a geographic coordinate system.

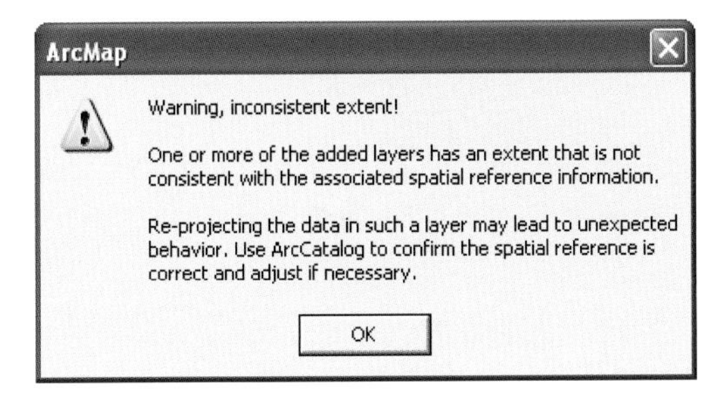

Figure 1–2 The coordinate system defined for the data does not match the spatial extent of the data.

The incorrect coordinate system definition must be removed from the data before proceeding. To do this:

- Delete the incorrect projection definition associated with the data by opening ArcToolbox > Data Management Tools > Projections and Transformations. (In ArcGIS Desktop 10, from within ArcMap, select Catalog window > Toolboxes > System Toolboxes to access this path.)
- Open the Define Projection tool, shown in figure 1–3, and from the Input Dataset or Feature Class drop-down list, select the name of the dataset that generated the error message.
- Click the Coordinate System browse button.
- On the Spatial Reference Properties dialog box that comes up, click Clear as shown in figure 1–4, then Apply and OK.
- Click OK again on the Define Projection tool to remove the incorrect coordinate system definition from the data.

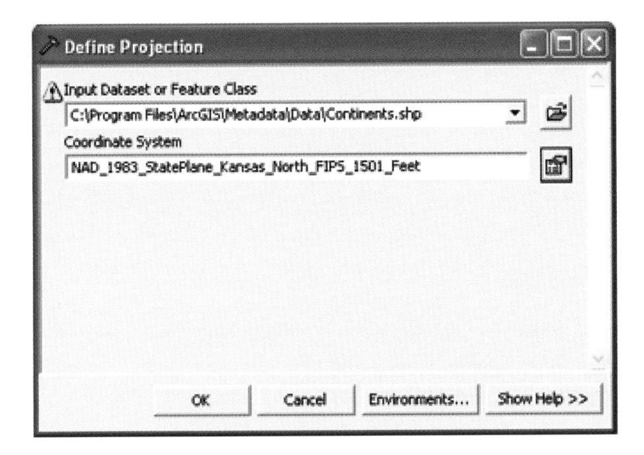

Figure 1–3 Define Projection dialog box. The shapefile specified in the Input Dataset or Feature Class box, Continents.shp, the continents of the world, should not have the coordinate system defined as NAD 1983 StatePlane for Kansas, or any state plane coordinate system. A projection from the state plane coordinate system applies only to a single state or portion of a state in the United States, and should not be used for data outside that specific area. Refer to figures 3–1 and 3–2 (pages 28-29) for illustrations of the areas covered by each state plane coordinate system FIPS zone.

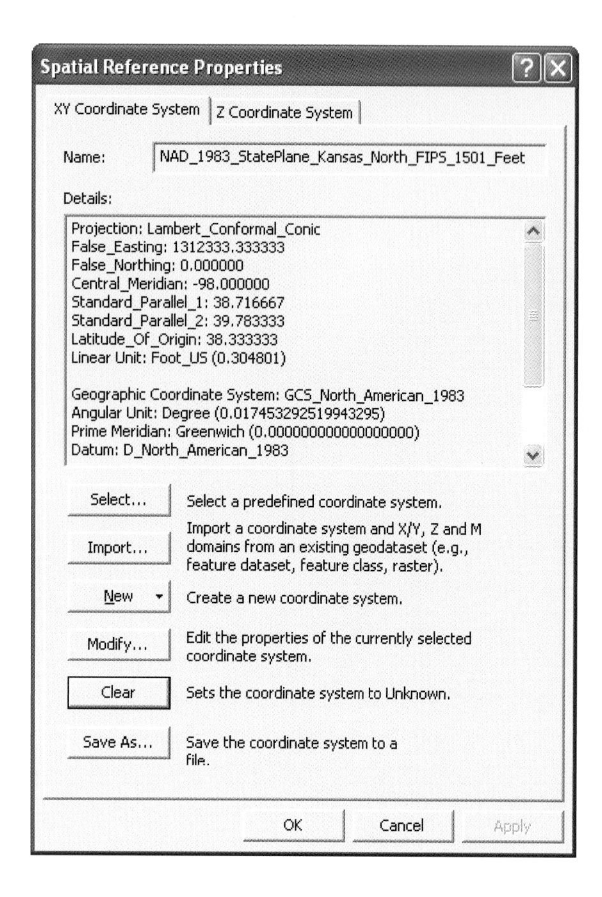

Figure 1-4 To clear the incorrect coordinate system definition from the data, click the Clear button, then click Apply and OK. The incorrect projection definition will be removed from the data.

After clearing the incorrect coordinate system definition from the data, you must also remove the incorrect coordinate system definition from the ArcMap data frame.

- In the top bar of the ArcMap window, click View > Data Frame Properties > Coordinate System tab.
- Click Clear in the upper right corner of the dialog box, then click Apply and OK. This removes the coordinate system from the data frame, and prepares for identifying the coordinate system of the data.

To identify the type of coordinate system for the dataset, proceed with the instructions that follow.

EXAMINING THE EXTENT OF THE DATA

Right-click the name of the layer in the ArcMap Table of Contents > Properties > Source tab (and follow along with figures 5 through 7). In the Extent box near the top, count the number of digits to the *left* of the decimal for Top, Bottom, Left, and Right. (Ignore any digits to the right of the decimal.) These numbers to the left of the decimal are the extent of the data on the earth, *in the coordinate system of the data*. These values are meaningful only in relation to the correct coordinate system definition.

The question marks following the Extent values substitute for an abbreviation for the units of measure for the coordinate system. ArcMap reads the units of measure from the projection file. Since the data does not have a coordinate system defined, there is no projection file associated with the data, and ArcMap is unable to read the units.

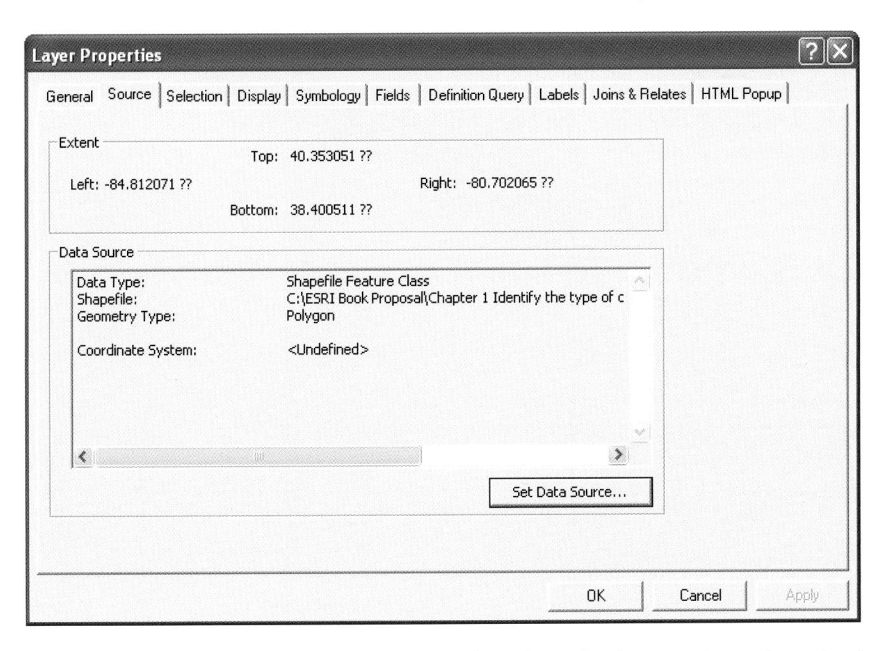

Figure 1–5 The two-digit numbers to the left of the decimal in the Extent box indicate this data is in a geographic coordinate system with units in decimal degrees. The question marks following the numbers are placeholders for the units. ArcMap reads the units from the coordinate system definition, but the projection is not defined, so ArcMap cannot display "dd," the abbreviation for decimal degrees. Refer to figure 1–10 to view the distribution of geographic coordinates across the globe.

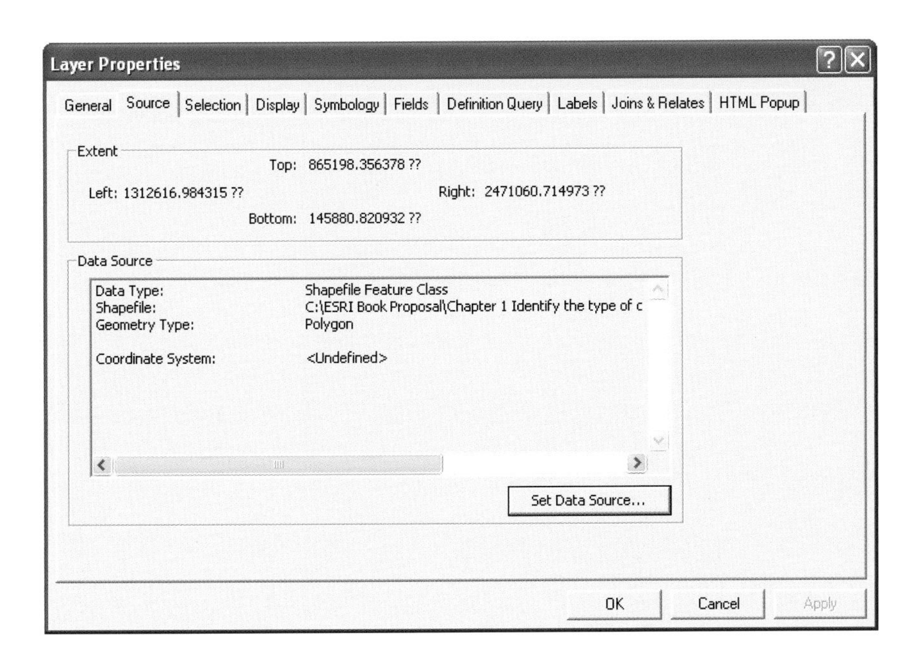

Figure 1–6 The units of measure for this dataset are also unknown, but we know the data is in a *projected* coordinate system because the numbers in the Extent box are 6 or 7 digits to the left of the decimal point. A projected coordinate system will generally have extent values 6 to 8 digits to the left of the decimal, although smaller or larger extent values can occur in some cases. The extent values are again followed by question marks. The coordinate system is undefined, so ArcMap cannot read the units of the projection.

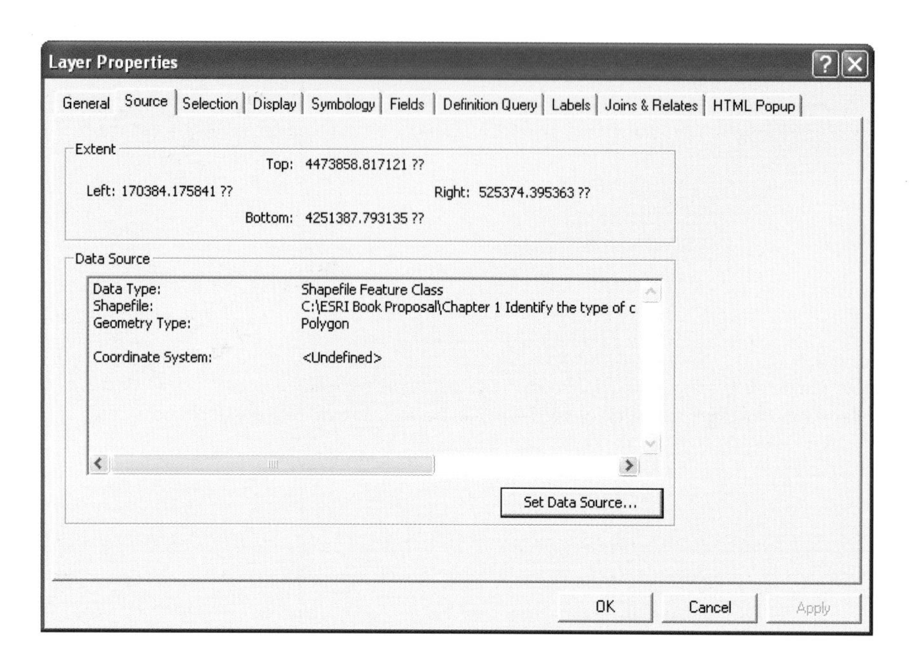

Figure 1–7 This dataset is also in a projected coordinate system, with extent values 6 or 7 digits to the left of the decimal. Comparing figure 1–6 with this screenshot, notice that the position of the number of digits is different. In figure 1–6, the Top and Bottom values are 6 digits to the left of the decimal. In this figure, the Top and Bottom values are 7 digits to the left of the decimal. In figure 1–6, the Left and Right values are 7 digits to the left of the decimal, while in figure 1–7 there are only 6 digits in those positions. This clue is important when working to identify the spatial reference for the data, as you will soon see.

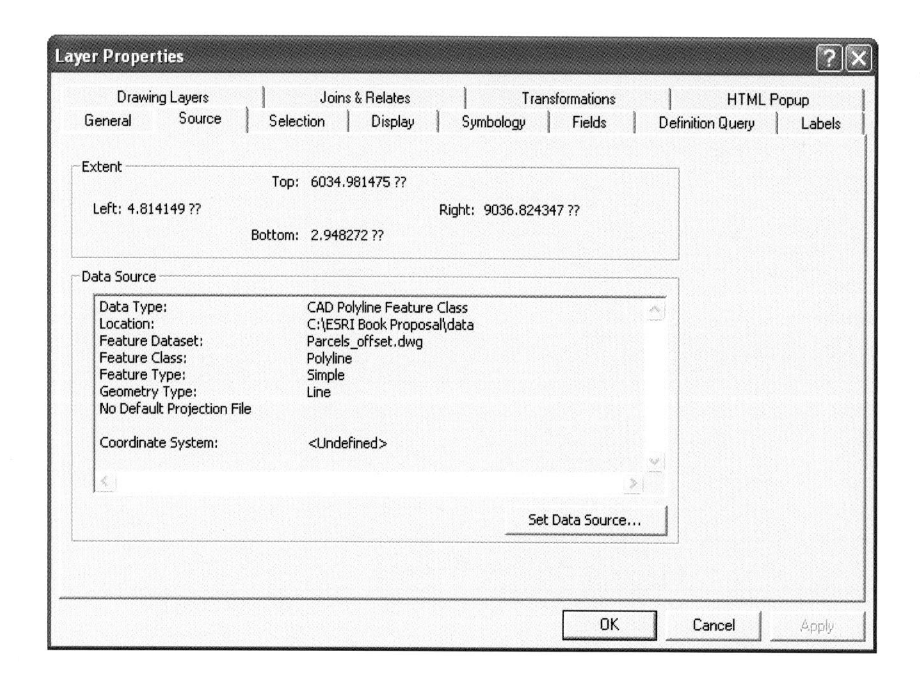

Figure 1–8 This AutoCAD drawing file (DWG) has an extent in a local coordinate system. The Left and Bottom coordinates are 1 digit to the left of the decimal, which could indicate data in a geographic coordinate system (GCS), but the Top and Right values are 4 digits to the left of the decimal, so this data has to be in a local coordinate system.

Note the number of digits to the left of the decimal on a piece of paper as a reference to be used in the following steps. You can make a quick diagram with the number of digits arranged as shown in figure 1-9, four samples of what the coordinates may look like for four given sets of data.

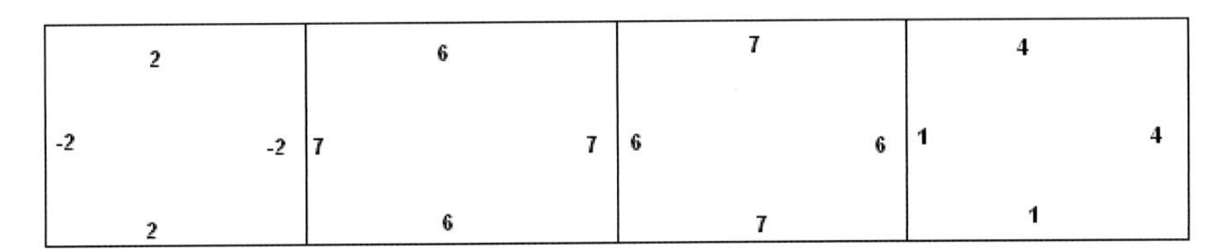

Figure 1-9 Diagrams of four sample extents for data in ArcMap, in dataset Layer Properties > Source tab > Extent box dialog. These numbers represent *only* the number of digits to the left of the decimal in the Top, Bottom, Left, and Right positions.

GEOGRAPHIC COORDINATE SYSTEM EXTENTS

In most cases, data that is in a geographic coordinate system (GCS) will have units of decimal degrees. A degree is an angle, and there are 360 degrees in a circle. In a GCS, the 360-degree extent is expressed in coordinates from -180° west to +180° east, measuring degrees of longitude or x-coordinates; and from +90° at the North Pole to -90° at the South Pole, measuring degrees of latitude or y-coordinates. These units are often referred to as "lat/long."

Within this coordinate extent, the location of data in decimal degrees will be expressed as positive or negative numbers 1, 2, or 3 digits to the left of the decimal for longitude, the Left and Right values. The latitude values, either positive or negative, can be no more than 1 or 2 digits to the left of the decimal for the Top and Bottom coordinates.

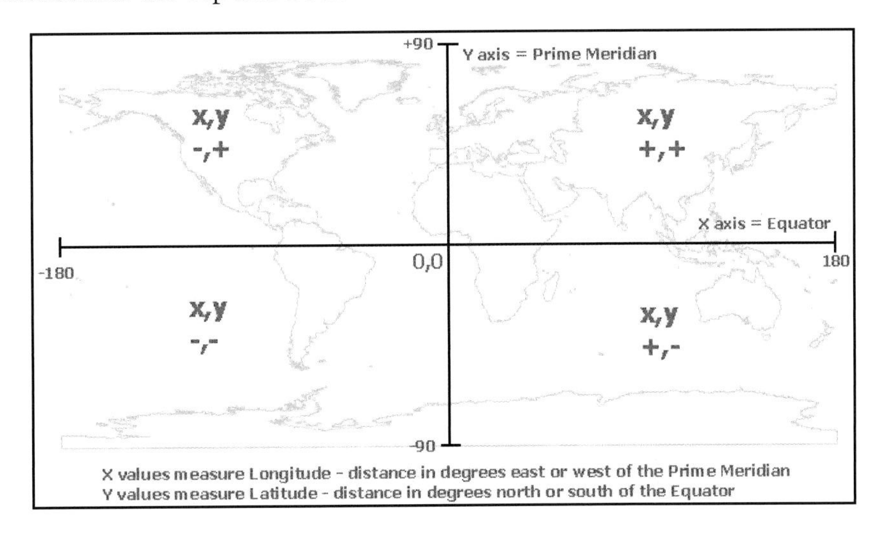

Figure 1-10 Distribution of geographic coordinates across the surface of the earth.

Data with coordinates in decimal degrees is in a GCS. This data can be created on hundreds of different datums. To define the coordinate system for data in a geographic coordinate system, the correct GCS and datum must be identified. (For detailed information about GCS and datums, refer to chapters 7 and 8.)

Identifying the correct GCS for the data

You have determined that your data has 1, 2, or 3 digit numbers to the left of the decimal, as viewed in the Layer Properties dialog box > Source tab in ArcMap, so the data is in a geographic coordinate system. Now you must identify the correct GCS for the data so that the coordinate system can be defined correctly.

You need to know which datums are typically used for the area in which this data is located. The most commonly used datums in the United States, for example, are North American Datum 1927 (NAD 1927), North American Datum 1983 (NAD 1983), World Geodetic System 1984 (WGS 1984), and North American 1983 (HARN). (HARN stands for High Accuracy Reference Network.) These datums are part of the Geographic Coordinate Systems named GCS_North_American_1927, GCS_North_American_1983, GCS_WGS_1984, and GCS_North_American_1983_HARN.

Hundreds of GCS and their associated datums for all parts of the world are supported in ArcGIS Desktop. If at all possible, request this critical information from the data source. If the data source is unable to provide this information, turn to chapter 2 for steps to identify the geographic coordinate system for the data.

PROJECTED COORDINATE SYSTEM EXTENTS

GIS and computer-aided design (CAD) data can be created using projected coordinate systems (PCS). Instead of angular units like decimal degrees, a PCS expresses the location of data using linear units that can be measured on the ground with a ruler. The most commonly used linear units are feet or meters, although other linear units can also be used. ArcGIS Desktop has a wide variety of predefined projection files already installed for PCS that apply to specific geographic areas, as well as to the entire world, using different map projections, coordinate systems, linear units, and datums. (Refer to chapter 10 for more information about projection files and projection parameters.)

Commonly used projected coordinate systems

In the United States, the most commonly used projected coordinate systems are state plane and universal transverse Mercator (UTM). Data projected to these coordinate systems, with units of feet or meters, will most often have extent values with 6 to 8 digits to the left of the decimal. As long as data projected to these coordinate systems falls completely within the area of the specified zone, the data will never display negative x or y values in the Extent box. Refer to figures 3-2 (page 29) and 3-7 (page 32) for illustrations of state plane and UTM zone extents.

In addition to these commonly used PCS, some states within the United States have created special coordinate systems for the entire state. These statewide projections are primarily created for states that are very large (Alaska, Texas), an odd shape (California, Florida, Michigan), or for distribution of statewide GIS datasets by the state (Georgia, Mississippi, Oregon, Wisconsin). These PCS are supported in ArcGIS Desktop, and should also be considered when identifying the projected coordinate system for these states. Data projected to these specialized state coordinate systems may display negative x or y values in the Layer Properties > Source tab > Extent box.

In addition to these PCS options, both Minnesota and Wisconsin have county coordinate systems for the state. Each county in these states has a specific coordinate system developed for use within that county. These county coordinate systems are also supported in ArcGIS Desktop.

If the data obtained is on a national or continental scale, data can be projected to coordinate systems created specifically for small-scale mapping. Some of these PCS are Albers equal area conic, equidistant conic, or Lambert conformal conic. Data in these coordinate systems may have negative x or y values in the data extent, and these coordinate systems should also be considered when working to identify the coordinate system for data in a PCS. (Turn to chapter 3 for steps to identify the projected coordinate system of the data.)

Small-scale vs. large-scale maps

Here the terminology of "small" and "large" is counterintuitive so it can be confusing.

Scale values can be thought of as ratios, and are unitless numbers. As an example, a scale of 1:100 means that 1 inch measured on the map is equal to 100 inches on the ground, 1 foot measured on the map equals 100 feet on the ground, and 1 meter measured on the map equals 100 meters on the ground.

Therefore, a *small*-scale map is one that displays data over a *large* area: a state like Alaska, an entire country like France, a continent, or the entire world. Scales shown in ArcMap may range from 1:1,000,000 to 1:750,000,000.

A *large*-scale map is one that displays data over a *small* area: a city, a county, a state plane FIPS zone (more about state plane FIPS zones later). Scales shown in ArcMap may range from 1:100 to 1:1,000,000.

LOCAL COORDINATE SYSTEM EXTENTS AND CAD

Data created using CAD software is frequently in a local coordinate system. A local coordinate system has its origin (0,0 or other values) in an arbitrary location that can be anywhere on the surface of the earth.

For example, when a new subdivision is planned, a surveyor will be hired to map out the parcels, streets, open space, and other land use within the subdivision. The surveyor will begin work at a point of origin that is a known location southwest of the subdivision. From that point, the surveyor will collect bearings and distances for lines that define the parcels and other features. Sometimes this point of origin is in a "real-world" coordinate system, such as state plane, but often it is assigned arbitrary local coordinates like 0,0 or other values.

Referring to what you see in the Layer Properties dialog box, the Extent of data created in a local coordinate system will most often have Top, Bottom, Left, and Right coordinates that are 3, 4, or 5 digits to the left of the decimal. In some cases, the Left and Bottom values may be 0 or other very small values, as illustrated in figure 1-8 (page 8).

If the CAD data has an Extent value that is 6, 7, or 8 digits to the left of the decimal, the CAD data was probably created in a real-world coordinate system such as state plane. (In this case, turn to chapter 3 and apply the techniques for identifying a standard projected coordinate system, before addressing the more complex processes of modifying or creating a custom projection file for the CAD data discussed in chapters 4, 5, and 6.)

CAD data is most often created with units of feet or meters, but other units are sometimes used and they can be unusual. There are CAD files created with units of centimeters, millimeters, kilometers, miles, inches, and in units of a tenth of an inch. In cases where the units of measure are very small, the values in the Extent box can have a very large number of digits (up to 14) to the left of the decimal, and still be in the state plane coordinate system. (The method for identifying and customizing units used in a standard coordinate system like state plane is addressed in chapter 4.)

CAD data in a local coordinate system can be aligned with other data in a projected coordinate system in ArcMap using one of four methods:

1. Modifying an existing projection file installed with ArcGIS Desktop.
2. Creating a custom projection file to align the data.
3. Transforming the CAD data in ArcMap.
4. Georeferencing the CAD data in ArcMap.

Instructions for options 3 and 4 above are provided in the ArcGIS Desktop Help. These alignment methods are not addressed in this book because the two methods only change the location where the CAD data is displayed in ArcMap. They do not provide for permanent alignment of the data.

Chapters 4, 5, and 6 address various options for aligning CAD data or other types of data files created in a local coordinate system. All of the methods outlined in those three chapters will apply to some CAD data files, depending on the spatial reference and production methods used by the data source.

SUMMARY

This book is about identifying the coordinate system for data, so that the coordinate system can be correctly defined and the data will draw in the right location in ArcMap in relation to other data. Because you may have issues to deal with immediately, chapter 1 starts you off with the information most likely to help you resolve them.

Data in a **geographic** coordinate system (GCS) most often has units in decimal degrees, which are angles. If the units of the dataset's coordinate system are in decimal degrees, the extent will have 1, 2, or 3 digit numbers to the left of the decimal, and some of these numbers may be negative. (Turn to chapter 2 for instructions on identifying the GCS for the data.)

Data in a **projected** coordinate system (PCS) most often has units in feet or meters. These are linear units that can be measured on the ground with a ruler. The extent will most often have 6, 7, or 8 digits to the left of the decimal. Most often these numbers will be positive, but they can also be negative. (Turn to chapter 3 for instructions on identifying the PCS for the data.)

Data in a **local** coordinate system most frequently has units of feet or meters, but other units might have been used to create the file. Usually CAD data in a local coordinate system has values in the Extent box with 3, 4, or 5 digits to the left of the decimal, although the Left and Bottom extent values in some cases may be 0 or other very small values. Coordinates can also be negative numbers. CAD data may also be created in a standard coordinate system, but the use of unusual units can cause extent numbers to be much larger (up to 14 digits) or smaller than usual.

Turn to chapter 4 for instructions on how to identify and modify a projection file to apply unusual units of measure. Chapter 5 offers instructions for modifying a standard projection file to align these data. Chapter 6 illustrates sample procedures that can be used to align rotated CAD files with other data.

CHAPTER 2

IDENTIFYING THE CORRECT GEOGRAPHIC COORDINATE SYSTEM

"How can I tell what datum my data is on?
I tried to find out from the data source, but they didn't know."

Let's say that chapter 1 helped you discover that your data is in a geographic coordinate system with units in decimal degrees—coordinate values in the Extent box (within the Layer Properties dialog box) are 1, 2, or 3 places to the left of the decimal. In a perfect world, the data source would provide you with the correct geographic coordinate system (GCS) and datum information for the data. In many cases though, either the data source will not have this information or the information provided will be incorrect. In those circumstances, you will need a procedure to identify the GCS of the data yourself. The process begins with obtaining accurate reference data with which to compare it.

In order to identify the GCS for your data, you require reference data that is precise and detailed, with a coordinate system already defined correctly, to use for comparison. The reference data must be in the same area on the earth's surface where the data with the unknown coordinate system is supposed to be located. You will compare the reference data with the unknown data in ArcMap to identify the correct GCS. You can use parcel data, street data, or aerial photography that is georeferenced with sufficient accuracy, or other detailed data available for this procedure.

You may find that the GCS for the two datasets, yours and the reference, do not line up because each is on a different datum. If so, a geographic transformation can be applied in ArcMap to align the two datasets.

If the offset (or displacement) between the data with the unknown GCS and the reference data is in a consistent direction and the offset is greater than 50 feet, then the offset is probably due to a datum shift between the data with the unknown GCS and the reference data. The offset between data on the NAD 1927 datum and the NAD 1983 datum is about 180 feet on the East Coast of the United States. Toward the central part of the United States, the offset between data on the NAD 1927 and NAD 1983 datums will decrease, and the offset will be primarily north to south by about fifty feet. The offset between data on these two datums increases again toward the western portion of the United States, and is about 300 feet on the West Coast. The offset can be as much as 450 feet in western Alaska.

Now we examine the offset that results when data on the NAD 1927 and NAD 1983 datums are drawn in ArcMap, and how the distance between features in the two datasets can be used to identify the unknown GCS.

EXAMPLE OF DATUM OFFSET BETWEEN NAD 1927 AND NAD 1983 DATUMS IN A GCS

The data used in the following examples represents features located in Riverside County, California. In this part of the United States, the offset between the NAD 1927 and NAD 1983 datums is approximately 300 feet. Assuming you have your own data, you can use it to follow along with the examples in this chapter. If you need data, see "Suggested data sources" below.

Suggested data sources

Street data with street names as attributes is available from many sources. StreetMap data is provided with ArcGIS Desktop on the ESRI Data & Maps CDs. Use of the detailed StreetMap data requires an additional license, but an evaluation license is available through your ESRI sales representative or through ESRI Customer Service.

National Agricultural Imagery Program imagery is available for download from the USDA's Geospatial Gateway at **http://datagateway.nrcs.usda.gov/**

ArcGIS Online is also a resource for imagery: **http://resources.esri.com/arcgisonlineservices/**

TIGER/Line data by county, including streets and street names, is available from the U.S. Bureau of the Census or can be downloaded in shapefile format, free of charge, from the ESRI Web site at
http://arcdata.esri.com/data /tiger2000/tiger_download.cfm
This data is in geographic coordinates, on the NAD 1983 datum, and must have the coordinate system defined before using the data as a reference layer in ArcMap. If using this data as reference, first use the Define Projection tool in ArcToolbox, then the Project tool to project the data to state plane or UTM, so the reference data will have linear units of measure applied. The accuracy of TIGER data varies substantially in different counties within the United States. It is very useful for providing street names, but other resources should also be used to obtain better data alignment.

To identify the correct GCS and datum on which your data with the unknown GCS was created, begin the process of comparison: Open ArcMap with a new, empty map. Click Add Data and add the reference data to ArcMap, then add the data that has no coordinate system defined. If necessary, change the symbols used to draw the data so that the unknown data and reference data are represented in contrasting colors. If using polygon data, to view and measure the offset between the polygon boundaries in the unknown data and the reference data more easily, you will need to change the polygon symbols from a filled symbol to hollow, then select contrasting outline colors. (If the Spatial Reference displays with the name GCS_Assumed_Geographic_1, refer to page 15, "How to remove an incorrect definition from the data.") If the reference data is also in a geographic coordinate system with units of decimal degrees, the data will align reasonably well, as shown in figure 2-1.

How to remove an incorrect projection definition from the data

If the Spatial Reference displays with the name **GCS_Assumed_Geographic_1**, this is *always* incorrect. GCS_Assumed_Geographic_1 is a default coordinate system assigned in earlier versions of ArcGIS Desktop to data with coordinates in decimal degrees (sometimes referred to as "lat/long") that did not have a coordinate system defined.

- If this coordinate system name appears for the data in ArcMap, open ArcToolbox > Data Management Tools > Projections and Transformations > Define Projection. (To access this path in ArcGIS Desktop 10, from within ArcMap, select Catalog window > Toolboxes > System Toolboxes.)
- From the Input Dataset or Feature Class menu, select the dataset that displays this incorrect coordinate system definition. Click the browse button for the coordinate system.
- In the Spatial Reference Properties dialog box, click Clear > Apply and OK, then OK again on the Define Projection tool to remove this incorrect coordinate system definition from the data.

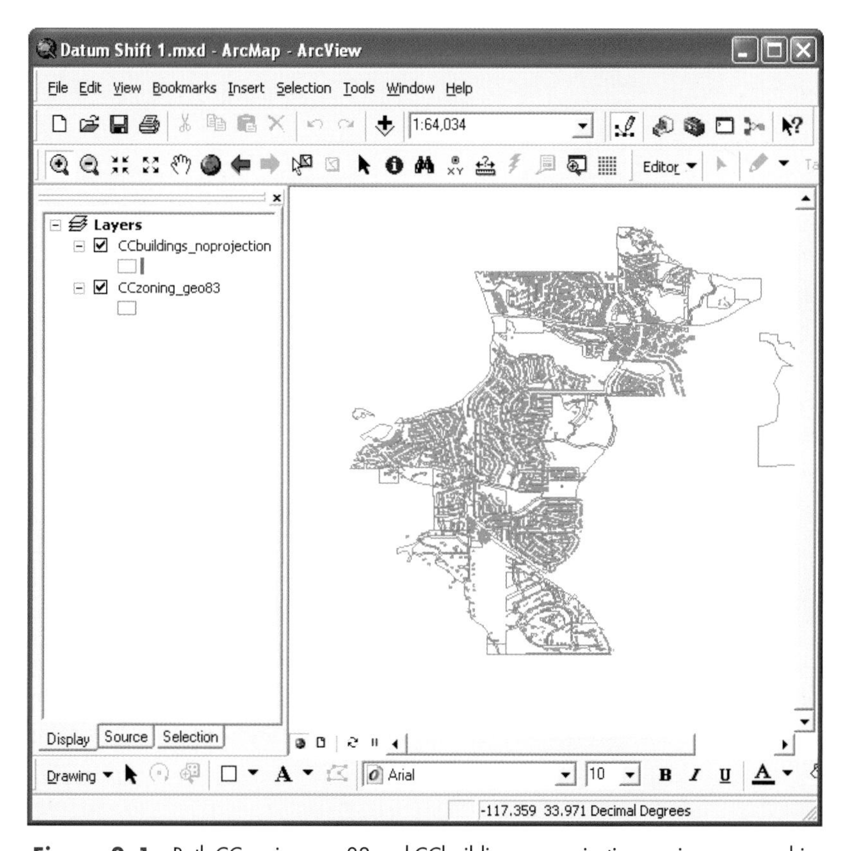

Figure 2–1 Both CCzoning_geo83 and CCbuildings_noprojection are in a geographic coordinate system. Note that the units of the coordinate system set for the data frame are Decimal Degrees in the status bar (lower right). The two datasets line up fairly well, but the buildings data layer is offset to the east (right) from the zoning layer. The GCS for the two datasets do not match.

Both sets of data are in a GCS, but are on different datums. You will need to identify the correct datum for the buildings layer. Go to View > Data Frame Properties > Coordinate System tab.

In the lower window labeled "Select a coordinate system:" open the Layers folder, as shown in figure 2-2, then open the folders labeled with the names of both datasets in the map.

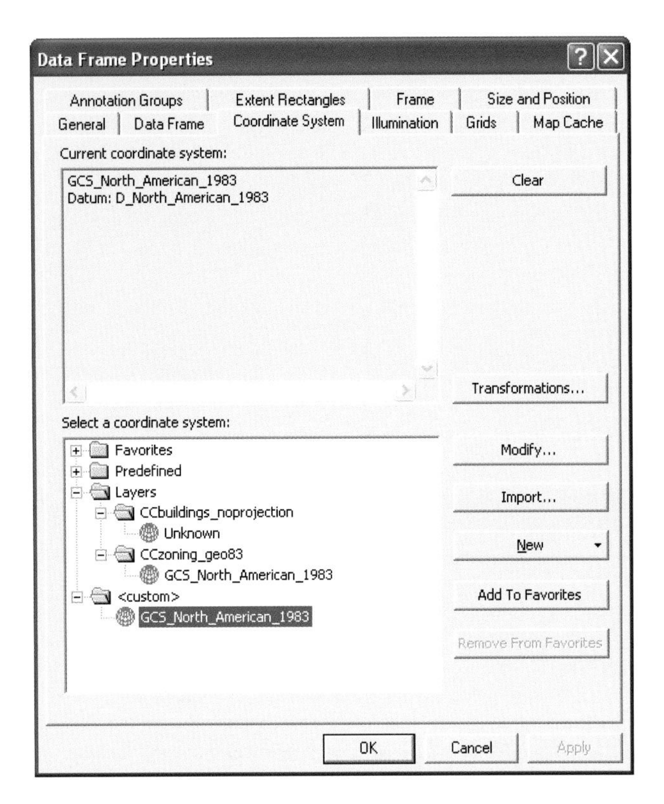

Figure 2-2 CCZoning_geo83 is really projected to geographic coordinates on the NAD 1983 datum. The coordinate system for CCbuildings_noprojection is not defined.

Let's think this through. The most common datum used for data within the United States is North American Datum 1983 (NAD 1983). If your buildings layer was on the NAD 1983 datum, however, the data would line up much more closely in the ArcMap window. Also, there would not be a consistent shift in one direction across the entire extent of the data. If the offset was due to bad data, the data would probably be offset in all different directions. You would not see a consistent shift in one direction, and the offsets would be much smaller than the 300 feet you see in figure 2-3.

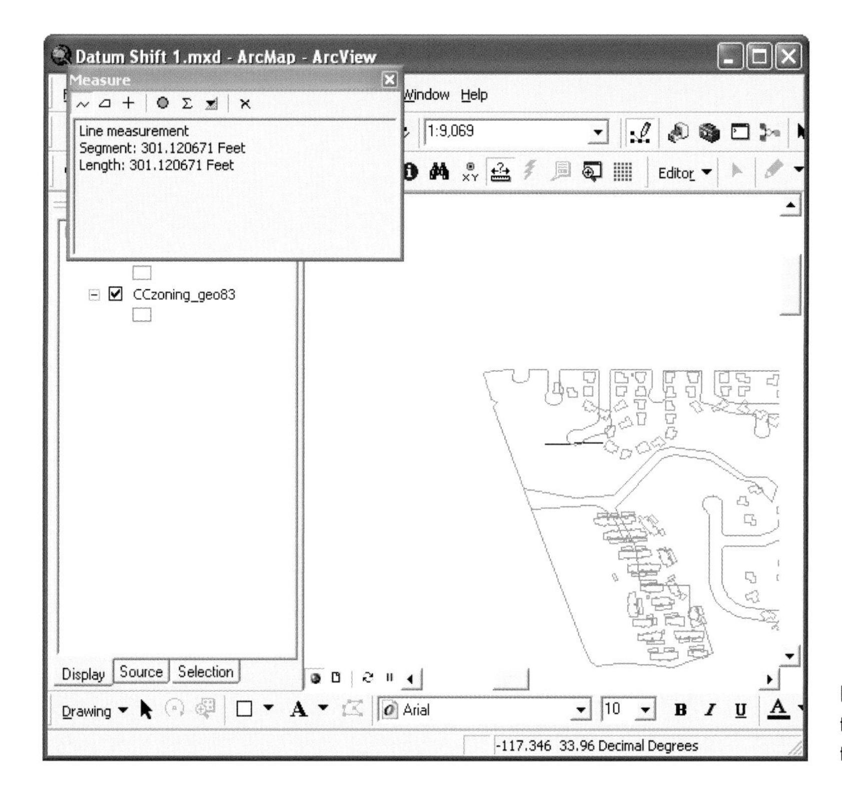

Figure 2-3 Zooming in, we see that the measured offset between the two datasets is about 300 feet.

Continuing our process of elimination, we know that another commonly used datum in the United States is WGS 1984. However, data on WGS 1984 and NAD 1983 are within one meter of each other. The offset is nearly 300 feet, so the buildings layer is not on the WGS 1984 datum either. The offset between NAD 1983 and NAD 1983 HARN (High Accuracy Reference Network; see page 103) is also quite small. Usually this offset is 1 meter or less, so CCbuildings_noprojection is not on NAD 1983 HARN.

The other datum fairly common in the United States is North American Datum 1927 (NAD 1927). Since data on the NAD 1927 datum is offset by a substantial amount from data on NAD 1983, check to see if the layer CC_buildings_noprojection is on this datum.

Under "Select a coordinate system" (as seen in figure 2-2), open the Predefined folder > Geographic Coordinate Systems > North America, and select the projection file named North American Datum 1927, shown near the bottom of figure 2-4.

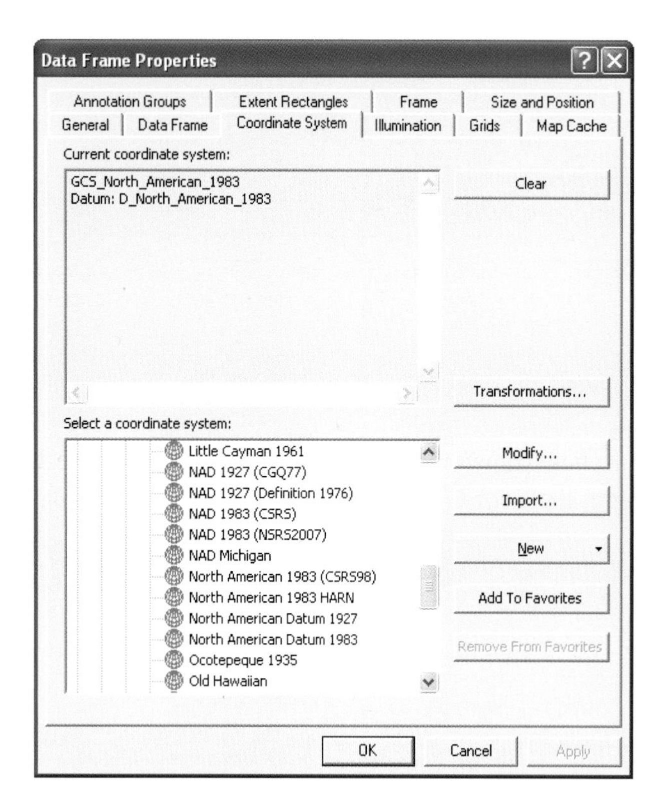

Figure 2-4 There are several choices for NAD 1927. The coordinate system North American Datum 1927 is used for data within the forty-eight contiguous states and Alaska. Refer to appendix A and the link to Knowledge Base article 29280 at `http://support.esri.com` to access "Geographic Coordinate Systems and Area of Use," a list you need to determine the area of use for all these different versions of NAD 1927.

After selecting this coordinate system definition, click Apply. The geographic coordinate system Warning dialog box will appear (figure 2-5).

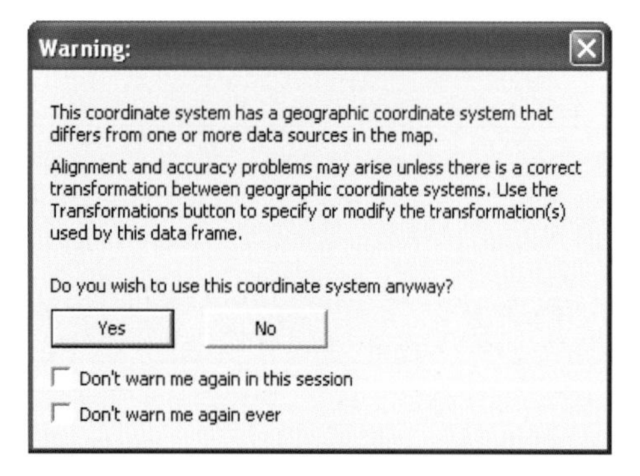

Figure 2-5 You must click Yes on the geographic coordinate system Warning dialog box in order to move on to set the correct geographic transformation for the ArcMap data frame.

A warning

In the Warning dialog box illustrated in figure 2–5, note the two check-boxes at the bottom, labeled "Don't warn me again in this session" and "Don't warn me again ever." *Under no circumstances should the bottom box ever be checked.* You need the warning to appear so you are notified that a geographic transformation is required because the data added is in a different GCS than the ArcMap data frame. As noted in the warning message, data alignment and accuracy problems will result if the correct geographic transformation is not applied to the ArcMap data frame.

After the geographic transformation is set for the ArcMap data frame as described in this section, the upper check box, "Don't warn me again in this session," may be checked. The current ArcMap session will continue to apply the selected geographic transformation until a new transformation is selected.

Click Yes on the Warning dialog box, then click Transformations (on the right side of the Data Frame Properties dialog box) to open the Geographic Coordinate System Transformations dialog box (pictured in figure 2–6).

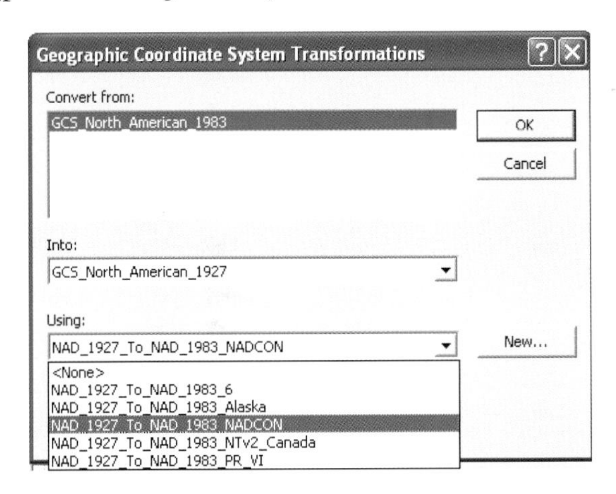

Figure 2–6 Select the correct geographic transformation for the area in which the data is located from the available options. Refer to appendix A and Knowledge Base article 21327. This article is linked to a list of supported geographic transformations and their areas of use. Look up these transformation names and their areas of use to determine the correct geographic transformation for your area. For data in the forty-eight contiguous United States, NAD_1927_To_NAD_1983_NADCON is the correct transformation to select.

At the top of the Geographic Coordinate System Transformations dialog box in figure 2–6 the Convert from box lists the geographic coordinate systems of the layers in the map document. The Into box displays the GCS of the ArcMap data frame—in this case, the GCS you set in figure 2–4. (The content of the Into box here must not be changed. To change it would defeat the purpose of the process you just accomplished in figure 2–4.) The Using drop-down lists any predefined geographic (datum) transformations between the two systems.

Datum transformation names: same name either direction

In figure 2–6 you see that you are transforming *from* GCS_North_American_1983 *to* GCS_North_American_1927, but the transformation names all specify NAD_1927_**To**_NAD_1983. This is because geographic transformations in ArcGIS Desktop are programmed to work in either direction. Regardless of whether you are transforming *from* NAD 1927 *to* NAD 1983 or *from* NAD 1983 *to* NAD 1927, you would select the geographic transformation with the same name.

To apply the selected transformation to the ArcMap data frame, select the transformation then click OK on the Data Frame Properties dialog box. When transforming between NAD 1927 and NAD 1983 within the forty-eight contiguous states, use the NADCON transformation. NADCON is the abbreviation for North American Datum Conversion. (For more information about this and other geographic transformation methods, see chapter 7.)

Click OK, then Apply, and OK on the Data Frame Properties dialog box.

The data has snapped into place, as shown in figure 2-7.

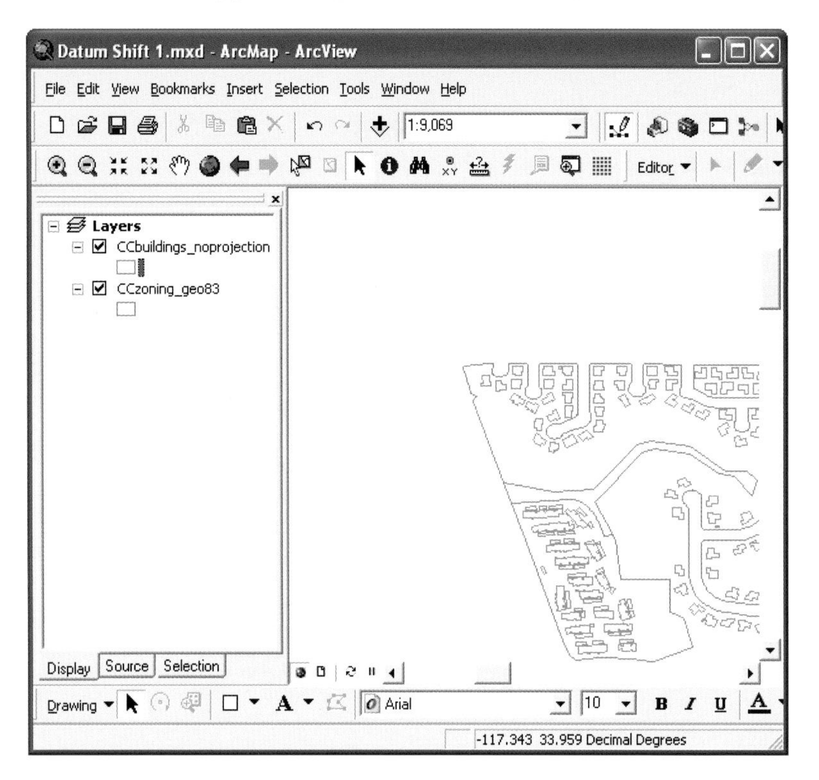

Figure 2-7 After applying the NAD_1927_To_NAD_1983_NADCON transformation to the ArcMap data frame, the data aligns correctly.

With the coordinate system identified for the buildings layer, open ArcToolbox > Data Management Tools > Projections and Transformations > Define Projection Tool. (To access this path in ArcGIS Desktop 10, from within ArcMap select Catalog window > Toolboxes > System Toolboxes.) Define the coordinate system for the unknown layer to the same coordinate system applied to the ArcMap data frame, in this case North American Datum 1927.prj. Now that the correct coordinate system definition has been applied to the data, and the correct geographic transformation applied in the ArcMap data frame, the two datasets line up properly.

EXAMPLE IN WHICH DATA WITH THE UNKNOWN COORDINATE SYSTEM IS IN A GCS, AND REFERENCE DATA IS IN A PROJECTED COORDINATE SYSTEM

If the reference data is in a projected coordinate system with units of feet or meters, the reference data will display a great distance from the unknown data. You may not even be able to see the data with the unknown coordinate system in the ArcMap data frame. Figure 2-8 displays an example of this common situation.

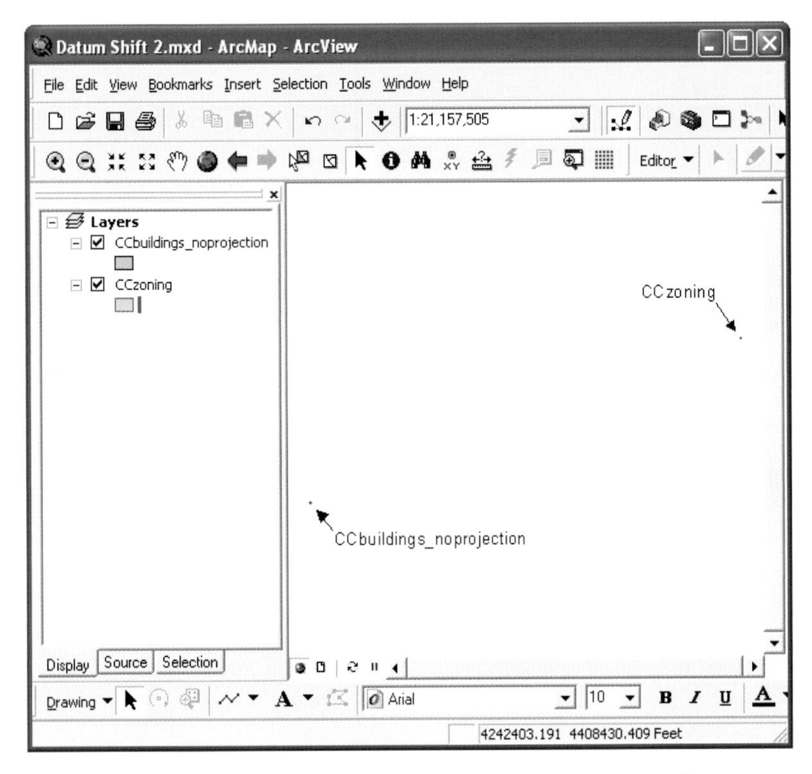

Figure 2-8 Because CCzoning is in a projected coordinate system with units of feet, and CCbuildings_noprojection does not have a coordinate system defined, the two datasets do not align at all. The buildings layer might be in a geographic coordinate system.

ArcMap cannot project the layer CCbuildings_noprojection on the fly because that dataset does not have a projection defined. Because the coordinate extent values of that layer are so much smaller than the extent values for CCzoning, and there is no coordinate system defined, ArcMap can only display CCbuildings_noprojection in relation to CCzoning based on the extent numbers. This places the buildings layer millions of feet away from CCzoning.

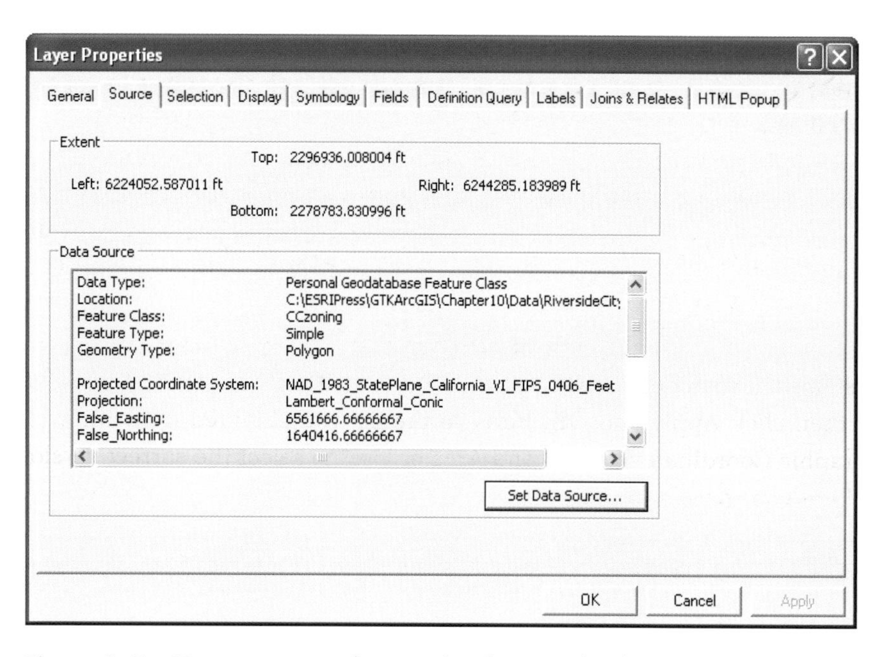

Figure 2-9 CCzoning is projected to state plane feet. Note that the *numbers* in the Extent box are all 7 digits to the left of the decimal, and all the values are positive numbers. This shows that the data in this layer is in a projected coordinate system.

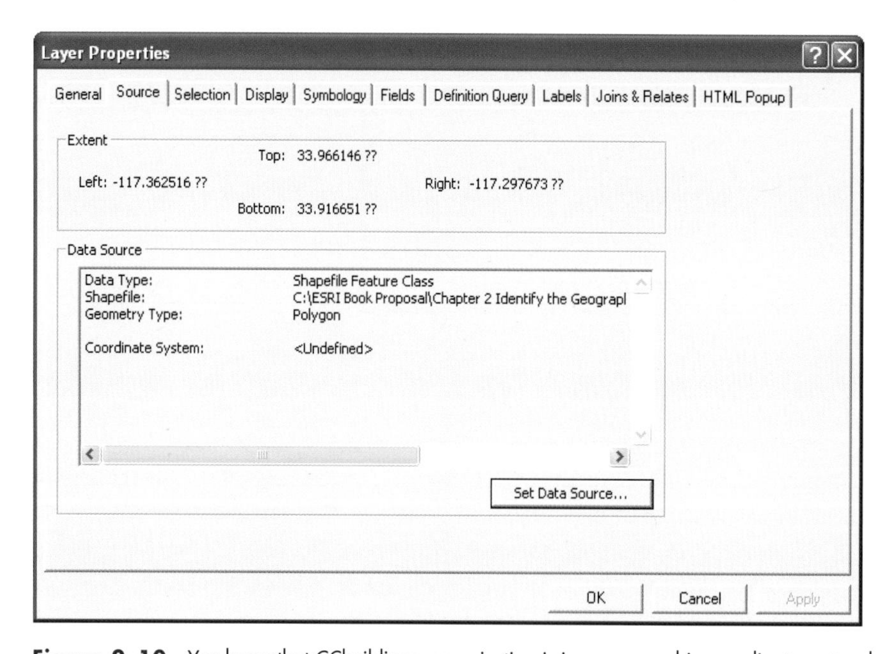

Figure 2-10 You know that CCbuildings_noprojection is in a geographic coordinate system based on the values in the Extent box, which are 2 or 3 digits to the left of the decimal. Notice that the Left and Right values are negative.

In the top bar of the ArcMap window, go to View > Data Frame Properties > Coordinate System tab. In the lower window on the coordinate system tab, open the folder labeled Layers, then open the folders for the two datasets in the map.

The datum name will be the first element in the coordinate system name for your reference data. If the data is located in the United States, the projection file name will probably begin with NAD 1983, NAD 1927, or NAD 1983 HARN. In this example, the CCzoning data is projected to NAD_1983_StatePlane_California_VI_FIPS_0406_Feet, so we will first apply the GCS, North American Datum 1983, to the ArcMap data frame.

Open the Predefined folder > Geographic Coordinate Systems > North America. Select the geographic coordinate system that matches the GCS of the reference data, NAD 1983 to start with, as shown in figure 2-11, then click Apply and OK. Refer to the list you obtained from Knowledge Base article 29280, "Geographic Coordinate Systems and Area of Use" to select the correct version of NAD 1983 for the area where your data is located.

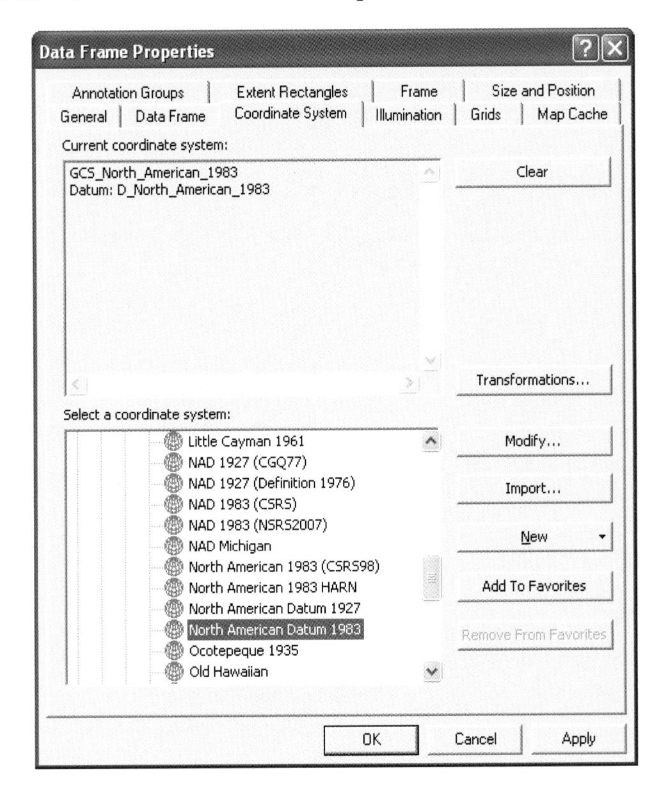

Figure 2-11 Select the geographic coordinate system that matches the GCS of the reference data. In this example, it is North American Datum 1983.

The reference data will project on the fly to the geographic coordinate system of the ArcMap data frame, and will align reasonably well with the unknown data. Zoom in closely to the data and examine the alignment of the unknown data with the reference data across the entire area. Figures 2-3 (page 17) and 2-12 show examples of the offsets you may see with your own data.

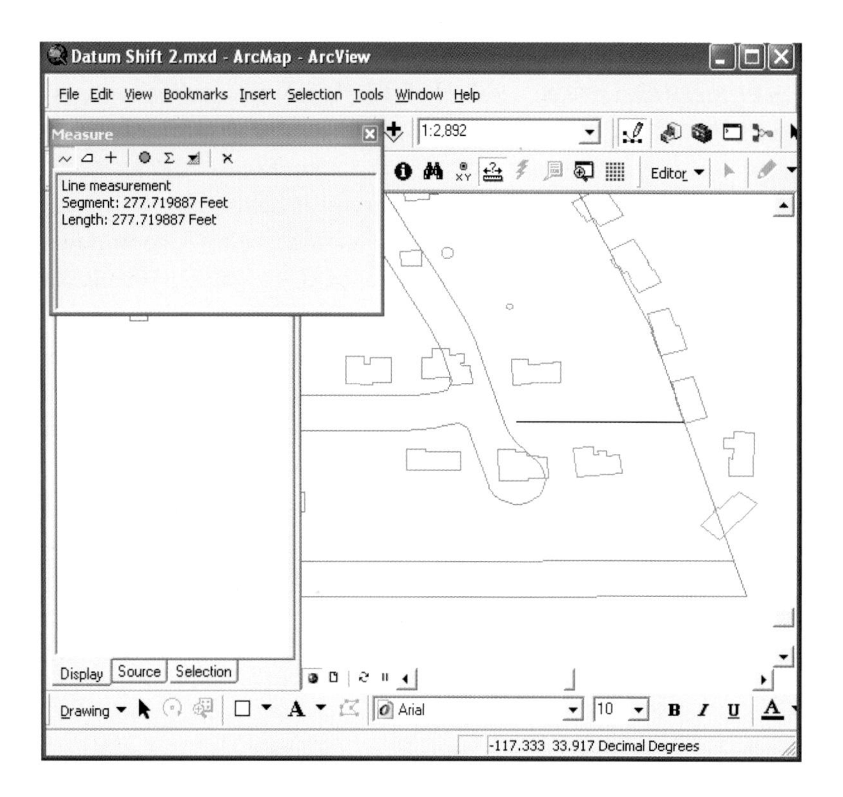

Figure 2-12 The buildings layer is offset from the zoning layer by nearly 300 feet. It is obvious that the buildings layer is not on the NAD 1983 datum.

In this example, we have buildings that are in the road. Noting an offset of almost 300 feet for this data in California, you can make the assumption that the data is on the NAD 1927 datum, not on NAD 1983.

Go back to View > Data Frame Properties > Coordinate System tab, and change the coordinate system of the ArcMap data frame to North American Datum 1927. Click Apply, then click Yes in the geographic coordinate system Warning box.

Click Transformations, and set the geographic transformation in the Using box to the correct transformation for this area, in this case NAD_1927_To_NAD_1983_NADCON.

Click OK, Apply, and OK. The data will align in ArcMap. Define the coordinate system for the data, using the Define Projection tool in ArcToolbox, as North American Datum 1927.

OTHER ISSUES TO CONSIDER

Is the offset between the unknown data and the reference data about the same distance all across the data? Does the data align exactly, or is it off just a few feet or meters? Use the Measure tool, set the distance units to feet or meters, and collect some exact measurements. Collect measurements across the entire extent of the data: the northwest (upper left); northeast (upper right); southwest (lower left); and southeast (lower right).

If the data is offset by only a few inches, feet, or meters, the offset may be due to other factors.

Is the ArcMap data frame set to North American 1983 HARN? HARN is a refinement of the NAD 1983 datum, used for survey-grade data. According to the National Geodetic Survey, HARN-grade data is supposed to be accurate to within +/- 1 centimeter. HARN stands for High Accuracy Reference Network, and the HARN datum is defined for limited areas on a state-by-state basis in most cases. The difference between HARN and NAD 1983 is at most a few feet. If the coordinate system of the ArcMap data frame is set to North American 1983 HARN, and the unknown data is on the standard NAD 1983 datum, a small consistent offset—in one direction across the entire extent of the unknown data—will be visible. The geographic (datum) transformation to align NAD 1983 data with data on the HARN datum for the correct state or area of the state must be selected in the Geographic Coordinate System Transformations dialog box. (Refer to chapter 8 for more detailed information, including screenshots of offsets between the NAD 1983 and NAD 1983 HARN datums.)

If the offset between the reference data and the unknown data is a few feet or meters, but the offset distance is inconsistent and the offsets are in different directions across the extent of the data, the problem is a data-quality issue. Inconsistencies between datasets that are due to data-quality issues should be reconciled. Doing so requires communication with the data provider, and may require additional surveying, ground inspection, or obtaining more reference data to determine which dataset is in the "right" place.

In this situation, the geographic coordinate system of the new data probably matches that of the reference data. When you have identified the coordinate system of the data, you must use the Define Projection tool to define the coordinate system for the data that has no projection definition. Then contact data sources or obtain additional reference data that will allow you to determine the correct location for features in the datasets.

DEFINING PROJECTION ERRORS AND WARNINGS

A warning message in green text, "datum conflict between map and output," may appear in the Define Projection dialog box if the GCS assigned to the data is different from the GCS of the ArcMap data frame. This message can be ignored if you have already set the geographic (datum) transformation for the ArcMap data frame. If the geographic transformation has not yet been set for the ArcMap data frame, refer to figures 2-5 and 2-6 (pages 18 and 19) and their accompanying text for instructions on how to apply the required geographic transformation to the ArcMap data frame.

The Define Projection tool may fail with the message "The dataset already has a coordinate system defined. Failed to execute." This tool failure occurs if the dataset is *read only*, or the data is in a location to which you do not have write access. Cancel from the Define Projection tool. If you have not already done so, copy the data to a folder on the local hard drive. Remove the data from ArcMap, save the MXD file to remove the record of the data from the map cache, then use Windows utilities to change the permissions on the local folder and the folder contents (instructions are provided on page 4). Add the local copy of the data back to ArcMap, and repeat the Define Projection process. If the process still fails with the same error, close the ArcMap session, and run the Define Projection tool from within a new ArcMap session or from within ArcCatalog.

TRANSFORMING BETWEEN THE NAD 1983 AND WGS 1984 DATUMS

The offset between NAD 1983 and the WGS 1984 datum within the forty-eight contiguous states overall is approximately 1 meter. The offset is greater in Alaska, Hawaii, and other active seismic areas because these transformations do not compensate for ground creep or larger ground movement during a major earthquake event.

Three geographic transformations are supported in ArcGIS Desktop for use in the United States:
NAD_1983_To_WGS_1984_1
NAD_1983_To_WGS_1984_4
NAD_1983_To_WGS_1984_5

NAD_1983_To_WGS_1984_1 applies to the entire North American continent. This transformation assumes that NAD 1983 and WGS 1984 are identical, and the transformation parameters are all zeroes. The published accuracy of this transformation is +/- 2 meters.

The transformation NAD_1983_To_WGS_1984_4 is an older transformation and should no longer be used. This transformation has been superseded by _5.

The published accuracy of the transformation NAD_1983_To_WGS_1984_5 is +/- 1 meter.

The accuracy of these transformations depends on the geographic area in which the data is located. Within the forty-eight contiguous states, either transformation _1 or _5 can be used and the use of either can be justified. The critical consideration is that once the choice is made, the selected transformation should always be used when transforming between NAD 1983 and WGS 1984, to maintain consistency in the GIS data for the organization.

SUMMARY

The first three chapters outline the steps toward identifying the coordinate system of data in ArcMap and defining the coordinate system to match the data, if the data has been created using a standard projection. Chapter 2 explores methods to identify the geographic coordinate system and datum on which data in a GCS, with units in decimal degrees, was created. (For further discussion about geographic coordinate systems and geographic transformations, refer to chapters 7 and 8.)

The next chapter examines the methods used to identify the coordinate system of data created in a projected coordinate system with linear units of feet or meters.

CHAPTER 3

IDENTIFYING THE PROJECTED COORDINATE SYSTEM

*"I received a parcel shapefile from the county,
but it doesn't line up with my other data in ArcMap. What now?"*

You determined in chapter 1 that the data with the unknown coordinate system is in a projected coordinate system (PCS), because when you examine the data in ArcMap there are 6, 7, or 8 digits to the left of the decimal — Top, Bottom, Left, and Right — in the Extent box on the Layer Properties > Source tab. Now you need to find out *which* PCS. It may help to know that in the United States, data in a PCS is *most often* projected to either the state plane or UTM (universal transverse Mercator) coordinate system. So let's discuss these coordinate system options first.

STATE PLANE COORDINATE SYSTEM

The state plane coordinate system (SPCS) was designed by the U.S. Coast and Geodetic Survey of the United States in the 1930s, in order to provide a standard for map projections within the United States. The SPCS provides mapping accuracy of 1:10,000 within the area of each zone.

ORIGINAL ZONES IN THE STATE PLANE COORDINATE SYSTEM

In creating the SPCS, larger states were divided into zones while small states were assigned to a single zone. In some cases, such as New England, several small states were grouped into a single zone. The zone boundaries were defined along state lines and almost always along county boundaries within the state. When originally defined, the SPCS was based on the NAD 1927 datum and Clarke 1866 spheroid.

Three different base projections were selected for these zones, depending on the shape of the zone. These projections were selected in order to minimize distortion of data within the zone. For zones that have an extent greater in the east-to-west direction, the Lambert conformal conic projection is used as the base projection. For zones that have an extent that is greater in the north-south direction, the transverse Mercator projection is used. Hotine oblique Mercator is a special case, used only for the Alaska Panhandle, because this zone lies at an angle, instead of being oriented either north-south or east-west.

Figure 3-1 shows the state plane coordinate system zones as they were originally defined on the NAD 1927 datum by the U.S. Coast and Geodetic Survey.

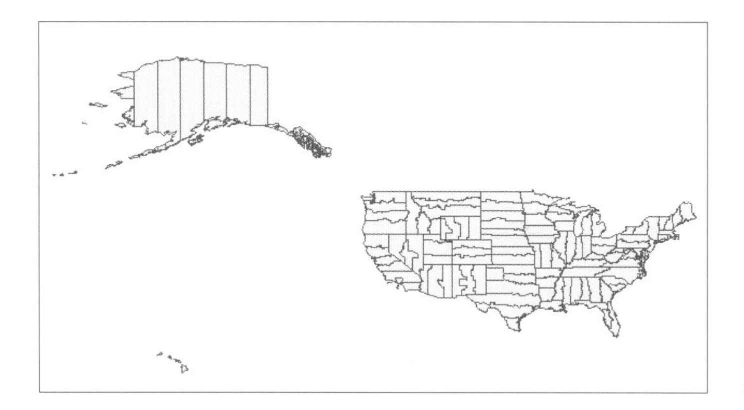

Figure 3-1 Original state plane coordinate system zones, defined on the NAD 1927 datum.

Viewing the zone shapes in the diagram, it is easy to see which base projection is used for each state plane coordinate system zone. The zones that are wider east to west are projected in Lambert conformal conic. Zones with a greater extent north to south are projected in transverse Mercator. The Alaska Panhandle, which lies at an angle, is projected in Hotine oblique Mercator. These projections minimize distortion in the areas thusly shaped.

Each zone has projection parameters calculated specifically for that geographic area. The SPCS projection for one zone should not be used for data in another zone in the same state or for data in another state.

FIPS ZONES IN THE STATE PLANE COORDINATE SYSTEM

The GRS 1980 spheroid was calculated from satellite measurements of the earth's surface, and the North American Datum 1983 is based on GRS 1980. Figure 3-2 illustrates the distribution of the state plane coordinate system FIPS zones, as defined on North American Datum 1983 (NAD 1983).

When the NAD 1983 datum was incorporated into the SPCS, the new keyword FIPS was used to designate each area. The acronym FIPS stands for Federal Information Processing Standard. Although the standard was never officially accepted, the zone numbers with the FIPS keyword are used regularly. What's important to remember is that the FIPS keyword, along with the associated FIPS zone number, is used in ArcGIS Desktop.

In ArcInfo Workstation, either ZONE or FIPS keywords can be used with the associated number, when defining the coordinate system for data projected to the state plane coordinate system. When ArcInfo coverage or grid data projected to the SPCS is added to ArcMap, and the coverage/grid has the coordinate system defined using the ZONE keyword, ArcMap automatically translates that zone number into the corresponding ArcGIS coordinate system and FIPS zone number.

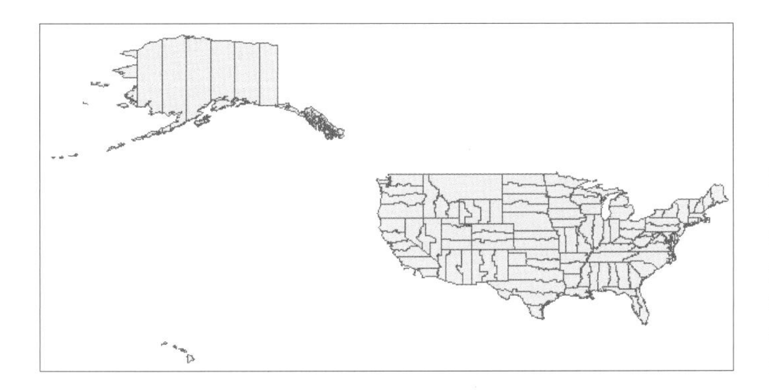

Figure 3-2 Updated state plane coordinate system zones, as defined on the NAD 1983 datum. Comparing these zones with those in figure 3-1, you can see evidence of the realignment of certain zone boundaries here: Montana has become one zone, for example, and Los Angeles County is now merging into California zone 5.

If the coordinate system of the unknown data is state plane, and the data lies in a zone that uses the Lambert conformal conic projection, Left and Right coordinate values will be larger numbers than the Top and Bottom, *unless the data is in eastern Texas*. Because the east-west extent of Texas is so large, in eastern Texas the Top and Bottom coordinate values will be larger than the Left and Right with the Lambert conformal conic projection. All other zones using the Lambert conformal conic projection will have larger extent values on the Left and Right.

If the data is projected to the SPCS, and lies in an area in which the base projection is transverse Mercator, the Top and Bottom coordinate values will almost always be larger numbers than the Left and Right.

The state of Florida (in figure 3-3) provides an excellent means of comparison between the SPCS base projections, because Florida is divided into zones that use Lambert conformal conic as well as transverse Mercator projections for different areas of the state.

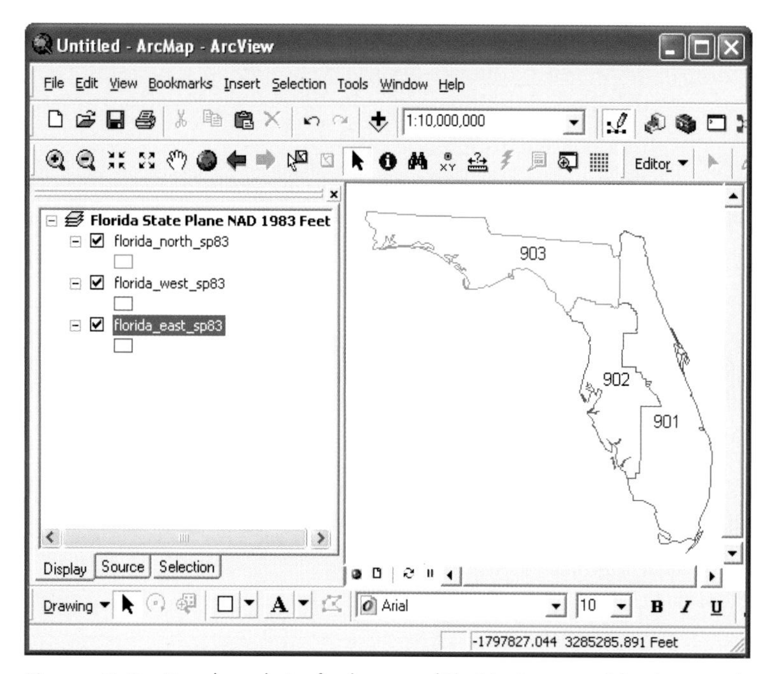

Figure 3-3 Zone boundaries for the state of Florida. Because of the shape and orientation of these areas, FIPSzones 901 and 902 are projected using transverse Mercator. FIPSzone 903 is projected using Lambert conformal conic.

Examining the shape of each zone, you can see that Florida North, FIPS 903, uses Lambert conformal conic as the base projection because the extent of the zone is greater east to west. Florida East and Florida West, 901 and 902, on the other hand, uses transverse Mercator as the base projection because these zones have a greater extent north to south.

Examine the coordinate extents for these three zones in figures 3-4, 3-5, and 3-6 and compare them.

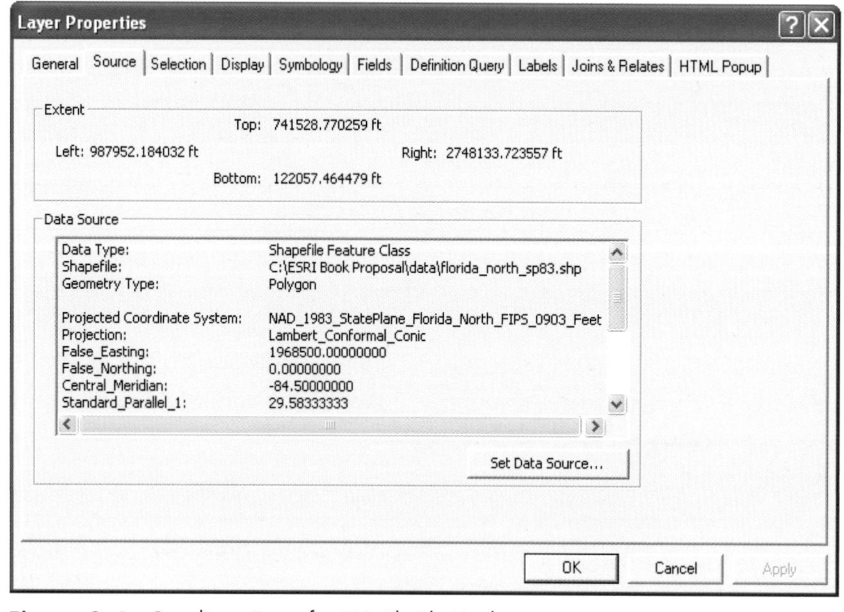

Figure 3-4 Coordinate Extent for 903, Florida North.

For Florida North (Figure 3-4), the Left and Right (longitude) coordinates are the larger numbers. Note that the projection is Lambert conformal conic.

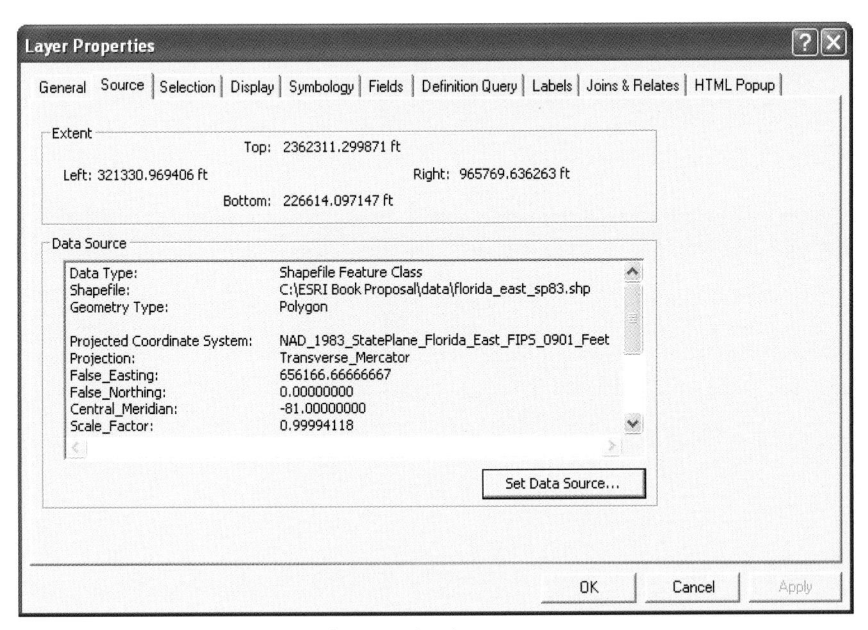

Figure 3-5 Coordinate extent for 901, Florida East.

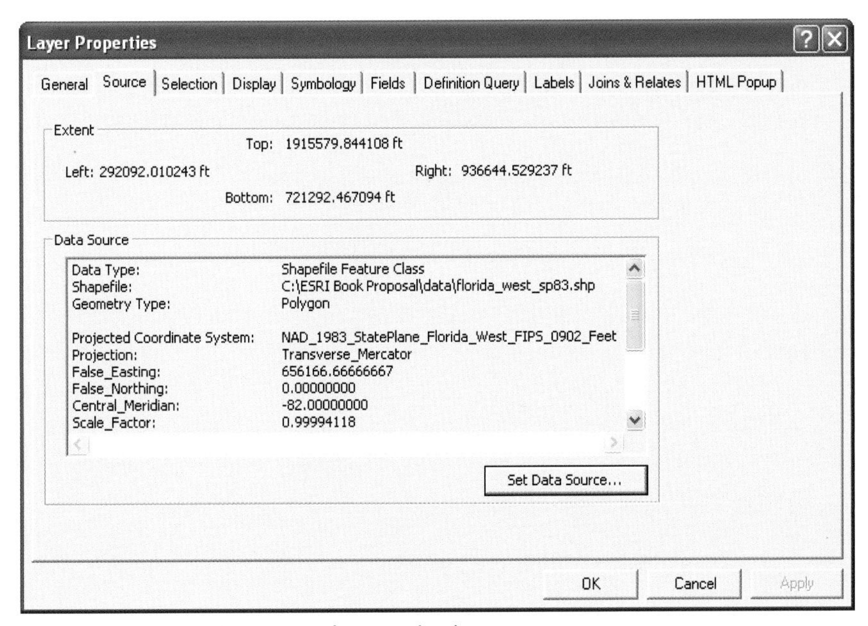

Figure 3-6 Coordinate extent for 902, Florida West.

For Florida East and Florida West, which are projected to the transverse Mercator projection, the Top coordinates are larger numbers than the Left and Right.

DATUMS USED WITH THE STATE PLANE COORDINATE SYSTEM

Four standard datums are available for use with the state plane coordinate system: NAD 1927, NAD 1983, NAD 1983 (HARN), NAD 1983 (NSRS2007). Additional datums are also supported for Hawaii and Guam, and other U.S. territories.

Three units of measure can be used with the state plane coordinate system: the U.S. survey foot, the international foot, and the meter. Projection files with these combinations of parameters are installed with ArcGIS Desktop.

Conversion factors

Units of U.S. survey feet (Foot_US) or international feet (Foot) can be used with the state plane coordinate system (SPCS). Certain states have legislated that the international foot can be used with the SPCS because the international foot converts exactly to a meter. The U.S. survey foot does not. Here are the conversion factors:

International foot (Foot) = 0.3048 of a meter

U.S. survey foot (Foot_US) = 1200/3937 of a meter or
= 0.3048006096012192024384048768097 5... .

UNIVERSAL TRANSVERSE MERCATOR COORDINATE SYSTEM

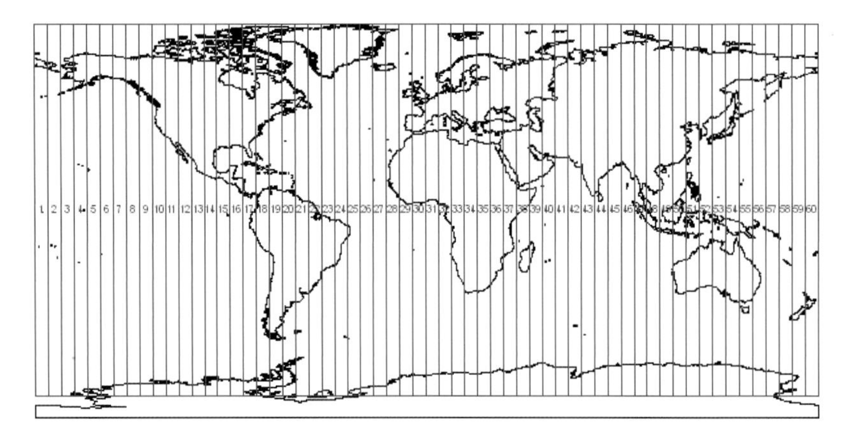

Figure 3-7 UTM zones of the world.

The UTM coordinate system divides most of the earth into sixty zones, each 6° wide. The UTM zones extend from 84° north latitude to 80° south latitude (-80°). This coordinate system was devised by the U.S. military, and the military specification for UTM includes meters as the standard unit of measure.

In the UTM projection files installed with ArcGIS Desktop, the UTM zones are designated by a number, followed by the letters N or S.

N refers to the area within the zone that is north of the equator.

S refers to the area within the zone that is south of the equator.

The UTM coordinate system also includes row designations. The rows run east-west and can measure 6 or 8 degrees in the north-south direction. Each row is assigned a letter designation. The row designations are used in the Military Grid Reference System (MGRS) graticule, and are *not* included in the projection file names in ArcGIS Desktop. The row designations are sometimes specified when collecting data with a GPS unit in the UTM coordinate system, but are not included in the projection file name. For example, when collecting GPS data for the state of Louisiana in UTM, the GPS unit would be set to collect data in UTM zone 15N for data north of the equator, not UTM zone 15 "S" for the row designation. Setting the GPS unit to collect data in UTM zone 15S would result in data displayed south of the equator in the Pacific Ocean when added to ArcMap.

Additional information and diagrams of the UTM coordinate system are available at
`http://earth-info.nga.mil/GandG/publications/tm8358.1/toc.html`

If data is in the UTM coordinate system with units of meters, Top and Bottom coordinates will always be 7 digits to the left of the decimal (unless near the equator and in the northern hemisphere). Left and Right coordinates will always be 6 digits to the left of the decimal.

EXTENT LIMITS OF UTM

The UTM coordinate system is based on the transverse Mercator projection. Because the transverse Mercator projection is calculated to minimize distortion in north-south trending areas, the east-west extent of data that can be projected using UTM is very limited.

The UTM coordinate system should not be used for any data that extends beyond the neighboring zones to the west or east, so the maximum east-west extent for data to be projected in UTM must not be greater than 18°. However, the distortion due to the projection increases a lot outside the base UTM zone. A UTM zone is designed to have no worse distortion than 1 part in 2,500 for data within the bounds of the zone. Compare this to the state plane's 1 part in 10,000. View this limitation in figure 3-8, which shows the extent of Texas east to west, in relation to the three UTM zones this large state crosses.

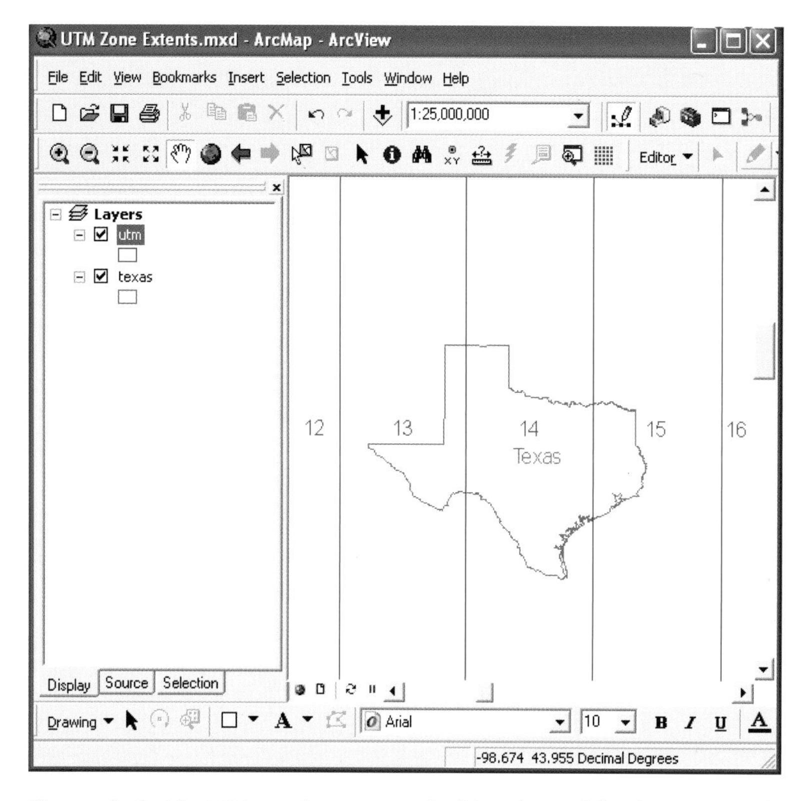

Figure 3-8 The UTM coordinate system should not be used for data covering the entire state of Texas. While the state fits within three UTM zones, the distortion due to the projection would be unacceptably high if statewide data were all projected to UTM zone 14N.

DATUMS USED WITH THE UTM COORDINATE SYSTEM

Because the UTM coordinate system can be used for data in all inhabited regions of the earth, this projected coordinate system can be used in conjunction with nearly any GCS. In the United States, UTM is most commonly used with the WGS 1984, NAD 1983, or NAD 1927 datums, but in other countries or areas of the world many other datums are also used.

A wide variety of projection files for UTM zones, which include GCS for specific countries or regions of the world, is installed with ArcGIS Desktop. You will need to refer to the file UTM.shp installed with ArcGIS Desktop to determine the UTM zone in which your data falls.

Refer to appendix B for the default installation location of ArcGIS Desktop for your version of the software and your computer operating system.

Projection files for the UTM coordinate system are installed with ArcGIS under **Coordinate Systems\Projected Coordinate Systems\UTM**.

UTM.shp is installed within ArcGIS in the folder named "Reference Systems."

In addition, refer to the list of geographic coordinate systems and their areas of use linked to Knowledge Base article 29280, as noted in appendix A, in order to determine the proper GCS for your data.

TESTING TO IDENTIFY THE PROJECTED COORDINATE SYSTEM FOR DATA

You can work through the available coordinate systems to find which one fits your data. If the data with the unknown coordinate system lies within the United States, begin by clicking Add Data, and navigate to the Reference Systems folder. (Refer to appendix B for the default installation location of ArcGIS Desktop.) Add the shapefile usstpln83.shp to ArcMap.

Right-click the name usstpln83.shp in the ArcMap Table of Contents and go to Properties > Source tab. Note that usstpln83.shp is in geographic coordinates with units of decimal degrees. The projection is defined as GCS_North_American_1983. The data with the unknown coordinate system and the newly added shapefile will not line up in ArcMap, and there will be a very large distance between the two data layers when zoomed to the full extent. (See page 21 for an illustration.)

Before you begin testing for the projections of data, set up the display properly. Because you need to be able to see the position of the data, set up the display so that the data you add to ArcMap doesn't cover the data you are trying to identify. Change the display properties of the data or the draw order of the files, so that the polygon shapefile usstpln83.shp does not cover your data.

Now think about the Extent numbers for your data. The unknown data has coordinates in the Extent box on the Layer Properties > Source tab that are all positive numbers, and the coordinates are most probably 6, 7, or 8 digits to the left of the decimal.

APPLYING STATE PLANE COORDINATE SYSTEM OPTIONS TO THE ARCMAP DATA FRAME

The state plane coordinate system is the most commonly used projected coordinate system in the United States. Therefore, it makes sense to test the SPCS options first, when working to identify the coordinate system for your data.

In ArcMap, click on View > Data Frame Properties > Coordinate System tab.

In the lower window labeled "Select a coordinate system:" open the folders Predefined > Projected > State Plane. Each of the folders in the State Plane directory (except international feet) will contain a projection file for each of the 121 SPCS zones. Select and apply the coordinate system for the state plane zone in which the data should be located from each of the folders as listed below:

NAD 1983 (feet)	units are U.S. survey feet
NAD 1983 (Intl. Feet)	units are in international feet
NAD 1983 HARN (Feet, Intl. and U.S.)	units may be in international or U.S. survey feet
NAD 1927	units may be in U.S. survey feet or meters
NAD 1983	units are in meters
NAD 1983 HARN	units are in meters
Other GCS	projection files for Hawaii and areas outside the fifty states

For example, the data used in figure 3-9 is supposed to be located in Riverside County, California. That county is located in California FIPS zone 0406. That specific FIPS zone will have a projection file installed in *each* of the above folders. The projection file from each folder will have the specified datum and units shown in the table above. If the projection file with a specific combination of datum and units of measure does not make the data draw in the correct location in relation to usstpln83.shp, open the next folder and try that datum/units combination.

After selecting a projection file, click Apply. If the geographic transformation warning shown in figure 2–5 appears, be sure to set the geographic transformation for the ArcMap data frame, click OK, then zoom to the full extent.

- If you are unable to see the data with no projection definition because it is too small in relation to the extent of the entire United States, right-click on the name of the dataset with the undefined projection, and select "Zoom to Layer."

- From the Drawing toolbar at the bottom of the ArcMap window, click the drop-down list next to the New Rectangle tool (the white square) and select the New Marker symbol from the bottom row.

- Left-click your mouse to drop a marker symbol in the middle of the data with the unknown coordinate system.

- Zoom to the full extent again. You will not be able to see the data, but you will see the marker symbol showing the location of the data in relation to the map of the United States.

Units tip: How to tell if the units from the selected projection file are right or wrong for the unknown data

If the data draws to the northeast of where the data should be, the linear units of the selected projection file are too big. If the data draws to the southwest of where the data should be, the units of the selected projection file are too small.

If the data was created with units of feet, but the selected projection file has units of meters, the data will draw a long distance to the northeast of where the data should be. The data will also be about three times too large, but at that scale you will not be able to see the size difference.

If the data was created with units of meters, but the selected projection file has units of feet, the data will draw a long way southwest of where the data should be. The data will also be only one-third the proper size, but you also won't be able to see the size difference at the current map scale.

The solution: select a different projection file, for the same zone with the same datum, but with different units. Refer to figure 3–9 for an illustration.

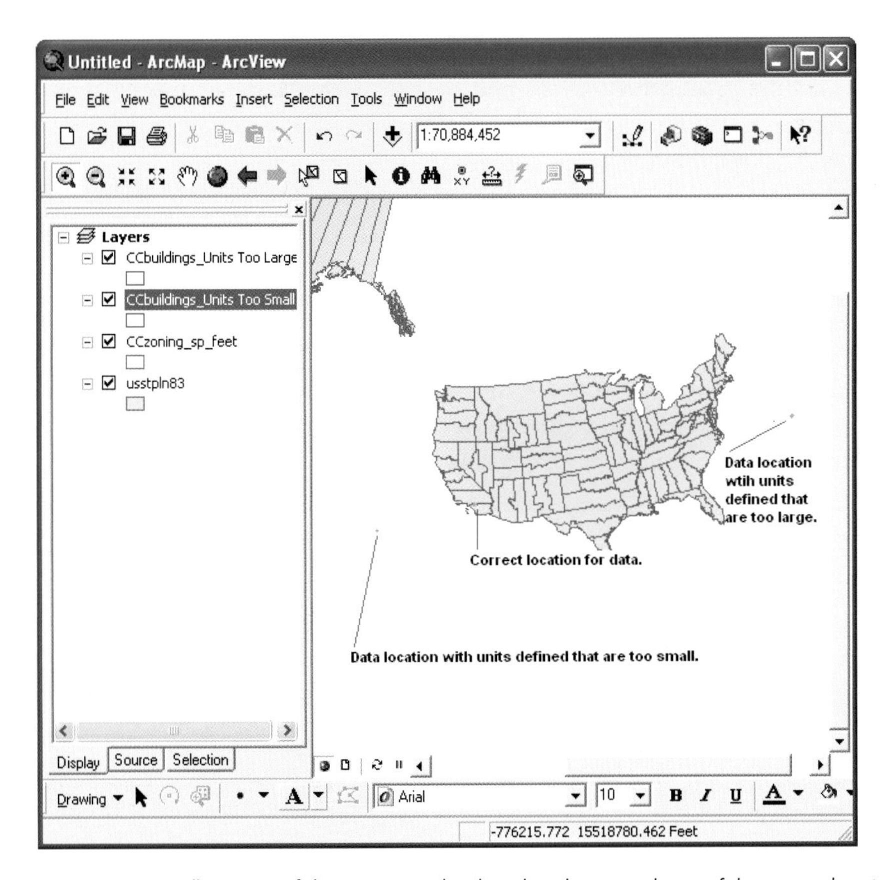

Figure 3–9 Illustration of the Units Tip. The data that draws northeast of the correct location is projected to the state plane coordinate system with units of feet, but has been *defined* with units of meters. The data southwest of the correct location is projected to the state plane coordinate system with units of meters, but has been *defined* with units of feet. Redefining the projections using the projection files that have the correct units will align the data. Turn to figure 4–1 (page 44) for an illustration of the units offset in the UTM coordinate system.

Datum tip: If the data is a long distance from where it should be, consider this:

For nearly all state plane zones, the difference between the false easting projection parameter defined on the NAD 1927 and NAD 1983 datums is *very* large.

For example, if the projection for the ArcMap data frame for FIPS 0401 (in the northern part of California) is set as NAD 1927 StatePlane Feet but the data is really projected to NAD 1983 StatePlane Feet, the data will draw in Wyoming instead of California. If the data is really projected to NAD 1927 StatePlane Feet but the projection of the data frame is set to NAD 1983 StatePlane Feet, the data will draw in the Pacific Ocean. You can immediately tell, therefore, if the datum selected for use with the SPCS and applied in the ArcMap data frame matches the data.

Watch out for North Carolina and South Carolina, though. The false easting values for the SPCS on the NAD 1927 and NAD 1983 datums for data in these two states are nearly identical. Therefore offsets for data in these states between NAD 1927 and NAD 1983 will only be about 180 feet, the difference of the datum shift for that part of the country.

Repeat assigning a state plane projection file from the available folders until usstpln83.shp snaps into place and the data with the unknown coordinate system appears in the correct area in the proper state. If the Geographic Coordinate System Warning dialog box appears, review the process for setting the geographic transformation in ArcMap outlined in chapter 2.

Verify the location by zooming in on the layer and using the Identify tool on the state where the data is drawn. Verify further by adding detailed reference data, such as parcels, streets, or other data that have projections correctly defined to the ArcMap document.

DEFINING THE COORDINATE SYSTEM FOR THE DATA AS STATE PLANE

If you have identified the projected coordinate system for the data as one of the state plane coordinate system options, note the name of the correct state plane coordinate system projection file and the location of the file in the directory structure.

- Open ArcToolbox > Data Management Tools > Projections and Transformations > Define Projection Tool. (To access this path in ArcGIS Desktop 10, from within ArcMap select Catalog window > Toolboxes > System Toolboxes.)

- From the "Input dataset or feature class" drop-down list, select the name of the dataset with the undefined coordinate system.

- To apply the coordinate system to the data, click on the browse button > Select > Projected Coordinate Systems > State Plane, and open the folder containing the correct coordinate system definition for the data.

- Select the correct projection file, click Add > Apply and OK > then OK again on the Define Projection tool. The selected coordinate system definition will be applied to the data.

IF STATE PLANE OPTIONS DO NOT ALIGN THE DATA, TESTING FOR UTM

If applying the various state plane coordinate system options does not align the data in ArcMap, examine the coordinate extent again:

Are the extent numbers all positive?

Are the extent numbers 7 digits to the left of the decimal on the Top and Bottom and 6 digits to the left of the decimal on the Left and Right? If these conditions are met, consider that the data may be projected to universal transverse Mercator (UTM) with units of meters.

Go back to View > Data Frame Properties > Coordinate System tab > Predefined > Projected coordinate systems > UTM folder. Select the correct UTM zone for the unknown data from the following datums:

NAD 1927
NAD 1983
WGS 1984

Coordinates in UTM meters on the NAD 1983 datum and coordinates for the same point on the WGS 1984 datum in the continental United States will be within a meter of each other.

Data in the UTM projection, on the NAD 1983 datum, is approximately 200 meters north of the same data on the NAD 1927 datum. There may be a slight shift either east or west between data on these two datums but an approximate 200-meter difference in the north-south direction is diagnostic. Actual offset distances between NAD 1927 and NAD 1983 coordinates may range from about 190 meters to 220 meters.

The 200-meter difference is comparatively slight; therefore, it is essential that precise comparison data be used to determine whether the correct datum has been selected. (For steps to set the geographic transformation in ArcMap, refer to chapter 2.)

What UTM zone to select?

- To determine which UTM Zone projection should be selected in ArcMap, click Insert > New Data Frame.
- Click Add Data, and add some of your own data, with the coordinate system correctly defined, to the new data frame.
- Click Add Data again, navigate to the Reference Systems folder, and add utm.shp to the map.
- Use the Identify tool to identify the zone number from utm.shp. UTM zone numbers for the fifty United States range from 4 for western Hawaii to 19 for eastern Maine. *The correct zone number is in the field named ZONE.* Ignore the values in the fields named ZONE_ and ZONE_ID. These fields were carried over from the original ArcInfo coverage, and these fields do not contain the correct UTM zone numbers.
- The new data frame can be deleted from the map document when the correct UTM zone number has been identified.

If you have identified the projected coordinate system for the unknown data as one of the UTM options, note the name and location of the correct UTM coordinate system projection file.

- Open ArcToolbox > Data Management Tools > Projections and Transformations > Define Projection Tool. (To access this path in ArcGIS Desktop 10, from within ArcMap select Catalog window > Toolboxes > System Toolboxes.)

- From the "Input dataset or feature class" drop-down list, select the name of the dataset with the undefined coordinate system.

- For the coordinate system, click on the browse button > Select > Projected Coordinate Systems > UTM, and open the folder with the correct datum name.

- Select the projection file for the correct UTM zone, click Add > Apply and OK > then OK again on the Define Projection tool. The selected coordinate system definition will be applied to the data.

OTHER PROJECTED COORDINATE SYSTEM OPTIONS

If neither the state plane nor UTM options align the data correctly and your unknown data has coordinate extents with values larger than 6, 7, or 8 digits to the left of the decimal, or if the extent coordinates include negative values, read on.

COUNTY COORDINATE SYSTEMS

In addition to using SPCS and UTM, the states of Minnesota and Wisconsin have developed specific county coordinate systems for each county in the state. GIS professionals in both these states are sensitive to these options, so will nearly always provide coordinate system information if data is in these county systems. These county coordinate system options must be kept in mind when working with data from either of these states.

STATEWIDE PROJECTED COORDINATE SYSTEMS

As mentioned earlier, some states have developed specific projected coordinate systems to project data for the entire state. This has been done for states such as Alaska, Texas, and California, where the area is very large, and both state plane and UTM coordinate systems split the state into multiple zones. (Remember that the state plane coordinate system should not be used for data outside the defined zone boundaries, while UTM should not be used for data more than 18° wide that extends past the neighboring zones to the west or east.) Other states that also use statewide projections are Florida, Georgia, Idaho, Michigan, Mississippi, Oregon, and Wisconsin. Data in these statewide projections may display negative coordinate values in the Extent box.

If your new data lies within these states, and did not align when either state plane or UTM options were investigated, test the statewide projections installed in ArcMap:

- Go to View > Data Frame Properties > Coordinate System tab.
- Open the Predefined folder > Projected Coordinate Systems > State Systems. These projection files may be on the NAD 1927, NAD 1983, or HARN datums. Some of the statewide projection files are provided with units of meters as well as feet.
- Remember to set the geographic transformation in ArcMap if the geographic coordinate system Warning dialog box (shown in figure 2–5, page 18) appears.

CONTINENTAL COORDINATE SYSTEM OPTIONS

Other projection files installed with ArcGIS Desktop have parameters for continent-wide datasets. Data projected to these coordinate systems may have negative values in the coordinate extents. If the data does not align using state plane, UTM, or specific state projections, these continental coordinate systems should also be tested.

To test these coordinate system options go to View > Data Frame Properties > Coordinate System tab > Predefined folder > Projected Coordinate Systems > North America.

If the unknown data is found to be in one of these predefined projection files, use the Define Projection tool in ArcToolbox to define the coordinate system appropriately. If none of the available projection files installed with ArcGIS Desktop aligns the data, refer to chapters 4, 5, and 6 for the guidance you need, whether instructions on modifying existing projection files to align data or instructions on creating custom projection files to meet this objective.

SUMMARY

Chapter 3 is predicated on the assumption that your data is in a projected coordinate system (PCS), but you do not yet know which PCS. This chapter outlines the process for identifying data in an unknown projected coordinate system and the techniques for identifying the units of the projection, using the project-on-the-fly utility in ArcMap. The chapter also provides suggestions for additional projected coordinate systems to check, if the standard options do not align the data.

All the options so far have been in feet or meters. The next chapter outlines techniques that you can apply in identifying units of measure used to create data that is not in feet or meters. Chapter 4 also includes instructions for modifying projection files installed with ArcGIS Desktop to accommodate other units of measure.

CHAPTER 4

WHEN STANDARD COORDINATE SYSTEMS DON'T WORK — WORKING WITH NONSTANDARD UNITS

"The CAD file my client sent me is huge when I add it to ArcMap with my other data. Why is that?"

When standard projection files installed with ArcGIS Desktop do not properly line up data in ArcMap, it is often because nonstandard units of measure were used in creating the data. Standard linear units — U.S. survey feet, international feet, and meters — are used in the projection files installed with ArcGIS Desktop, but other units of measure are often used when creating data. This is particularly true with data created using computer-aided design (CAD) programs.

You can, however, customize the linear units of the projection files installed with ArcGIS Desktop to match the data created with nonstandard units of measure. Modified this way and saved with a new name in ArcMap, the customized projection files can then be used to define the coordinate system for the data.

CUSTOMIZING THE UTM COORDINATE SYSTEM WITH UNITS OF FEET

In designing the universal transverse Mercator (UTM) coordinate system, the U.S. military specified units of meters. However, data in the United States is often collected and stored using units of feet. Therefore, customizing the UTM projection files (PRJs) installed with ArcGIS in order to use units of feet is a common practice.

From the units tip box on page 36, you will recall that when the coordinate system assigned to the data has units that are too large, the data will draw to the northeast of the proper location for the data. If a standard UTM coordinate system definition, with units of meters, has been assigned to the ArcMap data frame when working to identify the coordinate system, but the data draws far to the north and slightly east of where it should be in relation to the reference data, the data may be in the UTM coordinate system but the units could be feet rather than meters (see figure 4-1).

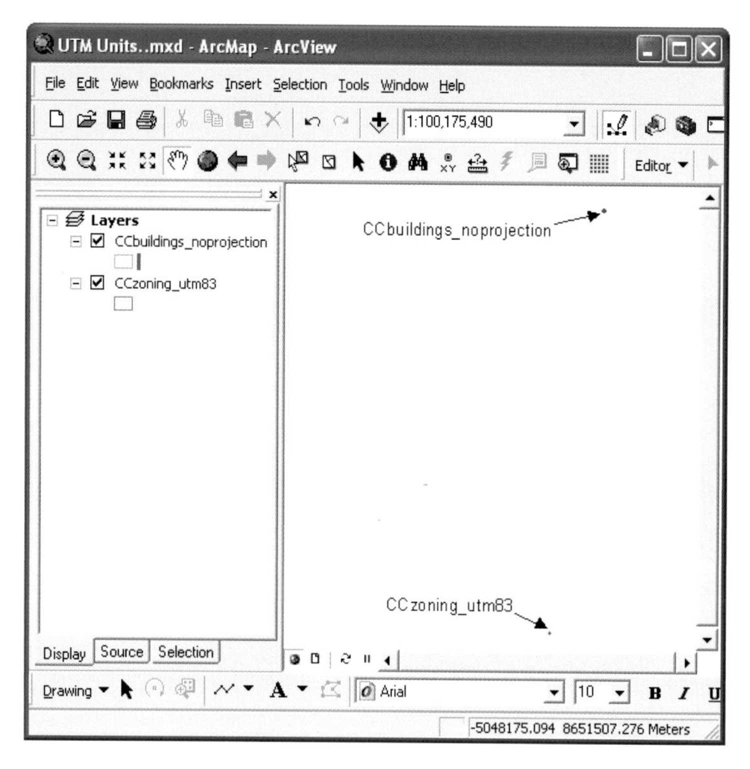

Figure 4-1 CCbuildings_noprojection appears north and slightly east of CCzoning_utm83, which is projected to NAD 1983 UTM zone 11N, with units of meters. This is another example of the offset that occurs when data in a projected coordinate system has a coordinate system assigned to it with units that are too large. Refer to figure 3-9 on page 37 for an illustration of this issue in the state plane coordinate system.

To modify the coordinate system definition applied to the ArcMap data frame and check if this is the issue affecting data alignment, go to View > Data Frame Properties > Coordinate System tab. At the top of the "Current coordinate system" window, the coordinate system must be set to the UTM Zone in which the unknown data is supposed to be located.

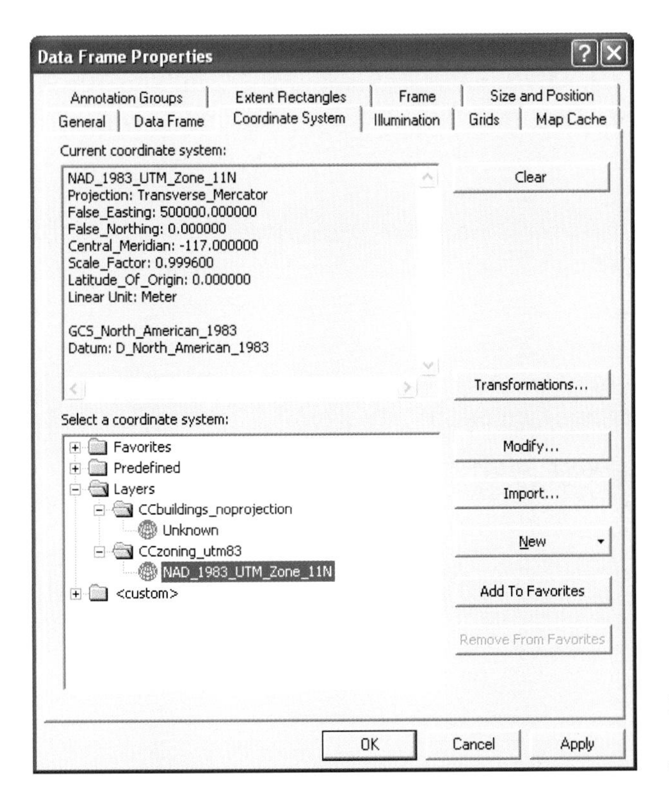

Figure 4-2 The ArcMap data frame is set to NAD 1983 UTM zone 11N, matching the coordinate system of CCzoning_utm83. Note that the Linear Unit is Meter.

On the right side of the Data Frame Properties > Coordinate System tab, click Modify.

In the Projected Coordinate System Properties dialog box, do the following:

1. Change the name of the projection file. The new name must not contain spaces or special characters, although underscores are permissible. When modifying a UTM projection file to change the units, changing the name indicates the presence of new units. You can simply add "_Feet", or "_Foot_US" at the end of the existing projection file name, as shown in figure 4-3. The new name for the projection file in this example is NAD_1983_UTM_Zone_11N_Feet.
2. Click on the Linear Unit drop-down list, scroll up to locate and select the unit "Foot_US" to apply to the custom projection file.

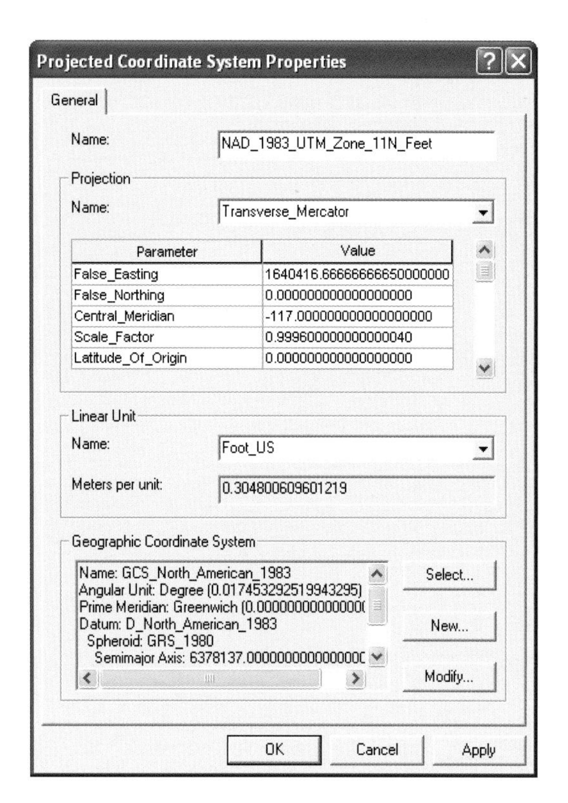

Figure 4-3 NAD 1983 UTM zone 11N projection file modified to use units of U.S. survey feet instead of meters. Notice that when the units are changed, the false easting of 500,000 meters is recalculated to 1,640,416.666, which is the false easting distance expressed in U.S. survey feet.

3. Click Apply and OK, then click Apply and OK on the Data Frame Properties dialog box. Examine the new location for the unknown data. Does the data now line up with the reference data? If the data lines up as in figure 4-4, the data is projected to UTM, but with units of U.S. survey feet (Foot_US).
4. If the data aligns correctly, using the custom UTM coordinate system with units of Foot_US, go back to View > Data Frame Properties > Coordinate System tab, and on the right, click Add to Favorites. This writes a copy of the custom projection file to disk, so that this file can be used to define the coordinate system for the data. (Refer to pages 59–60 later in this chapter for complete instructions on saving the custom projection file to disk and how to define the projection using the custom projection file.)

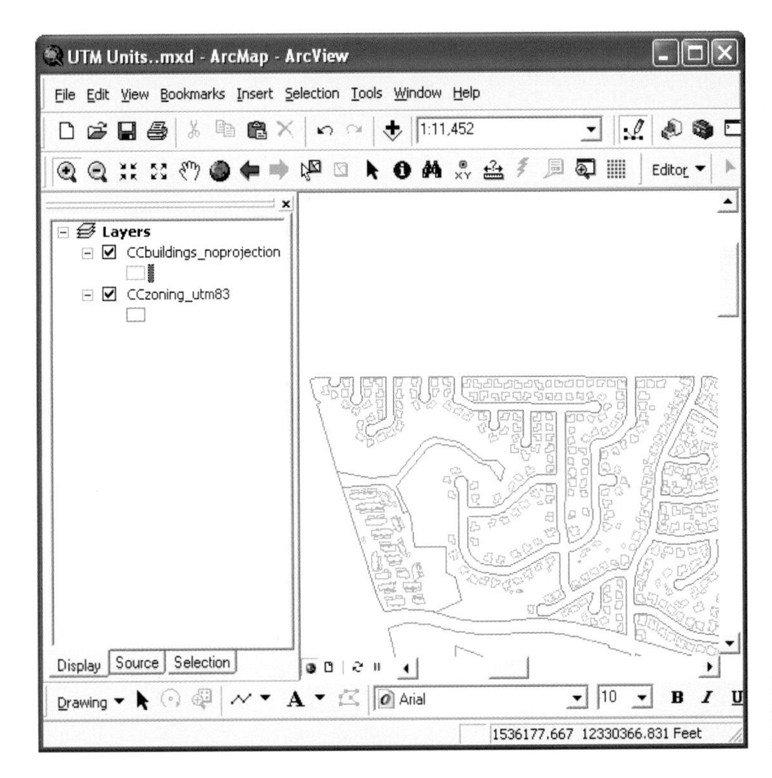

Figure 4-4 After modifying the projection file NAD 1983 UTM zone 11N to use units of Foot_US, the data aligns correctly in ArcMap.

DATUM OFFSET BETWEEN NAD 1927 TO NAD 1983 IN UTM

If you see a north-south shift of about 200 meters or 650 feet between the datasets, the unknown data may be on NAD 1927 instead of NAD 1983. If the datum for the projection file you just modified is NAD 1983, and the data draws south of where it should be (as shown in figure 4-5), change the projection of the ArcMap data frame to NAD 1927 for the proper UTM Zone, and modify that projection definition to units of feet. Remember to change the name of the projection file to reflect the change to units of feet.

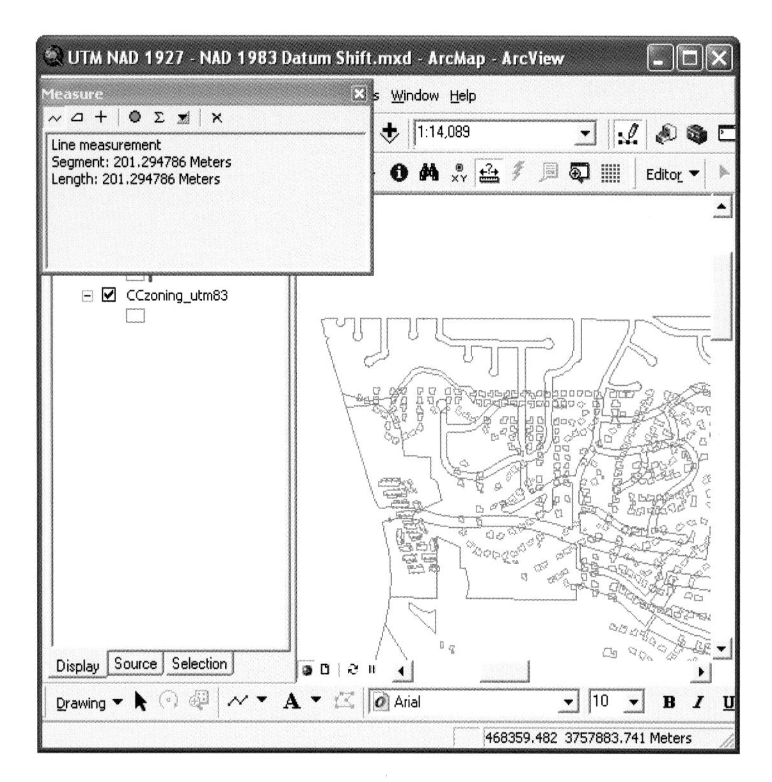

Figure 4-5 Offset, measured in meters, between NAD 1983 UTM and NAD 1927 UTM coordinates in Southern California.

When you click Apply on the Data Frame Properties dialog box, the geographic coordinate system Warning box will appear. Click Yes on the warning, then go to View > Data Frame Properties > Coordinate System tab.

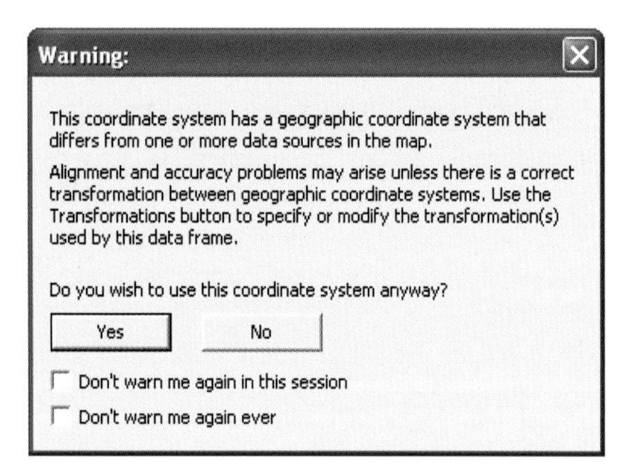

Figure 4-6 Sample of the geographic coordinate system Warning dialog box.

On the right side of the Coordinate System tab, click Transformations. A screenshot of the dialog that opens is shown in figure 2-6 (page 19). In the Geographic Coordinate System Transformations dialog box, the "Convert from" box lists the geographic coordinate systems of the layers in the map document. The Into box displays the GCS of the ArcMap data frame. The Using drop-down lists any available geographic (datum) transformations between the two GCS.

The content of the Into box in this dialog must *not* be changed. The Into box displays the GCS set for the data frame of ArcMap as demonstrated in figure 4-2. Changing the GCS in the Into box will defeat the purpose of this process. In the "Convert from" box, select the GCS that is different from the one shown in the Into box.

Click on the Using drop-down list, and select the correct geographic (datum) transformation for the area in which your data is located.

For data within the 48 contiguous states, select the transformation name NAD_1927_To_NAD_1983_NADCON. For data in Alaska, select the transformation name NAD_1927_To_NAD_1983_Alaska.

Old Hawaiian datum

Hawaii was never defined on the NAD 1927 datum. Originally, each major island of Hawaii was defined on a different datum, and these datums were collectively grouped as the "old Hawaiian datum." The transformation Old_Hawaiian_To_NAD_1983 is provided in ArcGIS Desktop to transform data for the state of Hawaii.

Remember that ArcMap applies geographic transformations in either direction. The same transformation name will be selected whether you are going *from* NAD 1927 *to* NAD 1983 or in the reverse direction, *from* NAD 1983 *to* NAD 1927.

Click OK on the Geographic Coordinate System Transformations dialog box, then Apply and OK on the Data Frame Properties dialog box.

If the data aligns correctly using the customized UTM projection file on the NAD 1927 datum with units of feet, refer to pages 59–60 later in this chapter for complete instructions on saving the custom projection file to disk and defining the projection for the data using the custom projection file.

CUSTOMIZING THE STATE PLANE COORDINATE SYSTEM WITH UNITS OF INCHES

CAD files are frequently created with units of inches or other small units to increase data precision. In this example, we examine the process of aligning an AutoCAD drawing file created with units of inches. Compare the size of Parcels.dwg with Index.dwg in figure 4-7. Index.dwg is projected to NAD 1983 StatePlane New Hampshire FIPS 2800 (Feet). Parcels.dwg is supposed to lie within Index.dwg, but in comparison with Index.dwg, Parcels.dwg is huge.

Parcels.dwg shows land ownership (red outlines), in addition to streets, trees, and other physical features, including the coastline where New Hampshire borders the Atlantic Ocean. Index.dwg primarily contains roads and subdivision boundaries, along with the coastline.

Supported CAD files

ESRI software, including ArcGIS Desktop, supports CAD files created with AutoDesk (AutoCAD) and Bentley (Microstation) software products.

AutoCAD saves files in drawing (DWG) or export (DXF) formats, and can save files in a variety of versions. DWG files are binary, while DXF files are text files. Both file formats store the same information, but DWG files store data more efficiently, so the DWG file will be smaller than the DXF file that contains the same data.

Microstation natively saves data in the DGN format. The DGN file is also binary. Microstation can also save files in the DWG or DXF format to add compatibility with AutoDesk products.

Information about support in ArcGIS Desktop for specific versions of AutoCAD and Microstation files is provided in the ArcGIS Desktop Help.

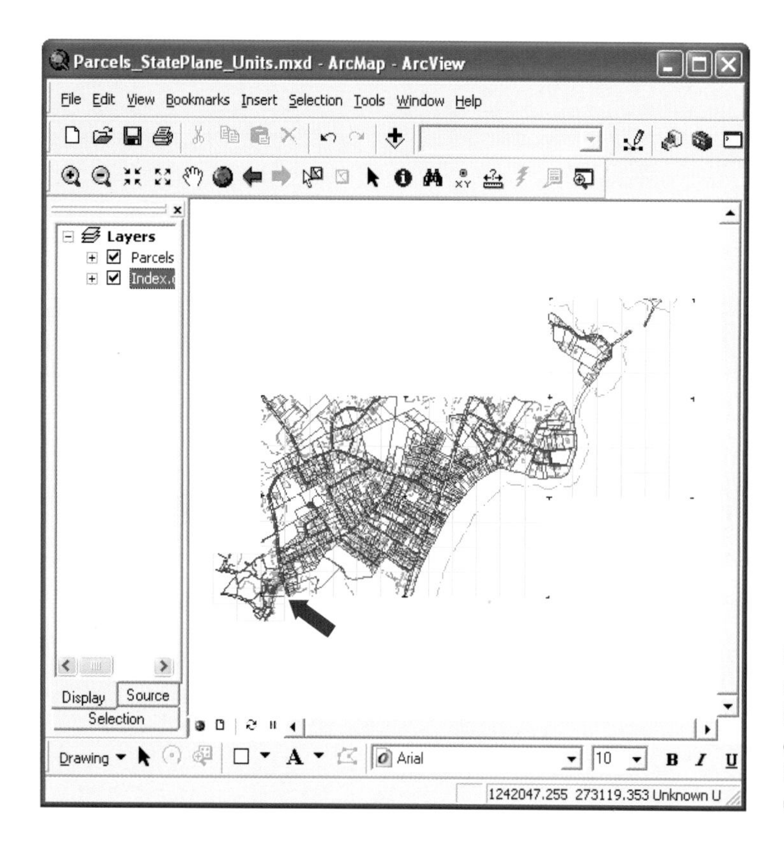

Figure 4-7 Compare the much larger size of Parcels.dwg with Index.dwg. The file Parcels.dwg should fit within Index.dwg (near bottom left) but appears much larger because the file was created with units of inches, not feet.

Based on the display of the data, it is obvious that the units for Parcels.dwg are not feet, and are much smaller units than those used for Index.dwg, the reference data.

In order to more easily identify features that exist in both drawing files, turn off some of the layers in Parcels.dwg, so that it is easier to locate features that exist in both datasets. Zooming in on Parcels.dwg and using the Identify tool, notice in figure 4-8 that the parcel lines drawn in red are on the layer named LOT-L. There is another layer called TREE-L, drawn with green lines. Buildings are drawn in cyan (Layer Name, BLDGS). These three layers are not useful for this process so turn these off.

Right-click the layer named Parcels, select Properties > Drawing Layers tab, and uncheck the boxes for BLDGS, LOT-L, and TREE-L, then click Apply and OK. Turning off these three layers makes the relationship between Parcels.dwg and Index.dwg much clearer.

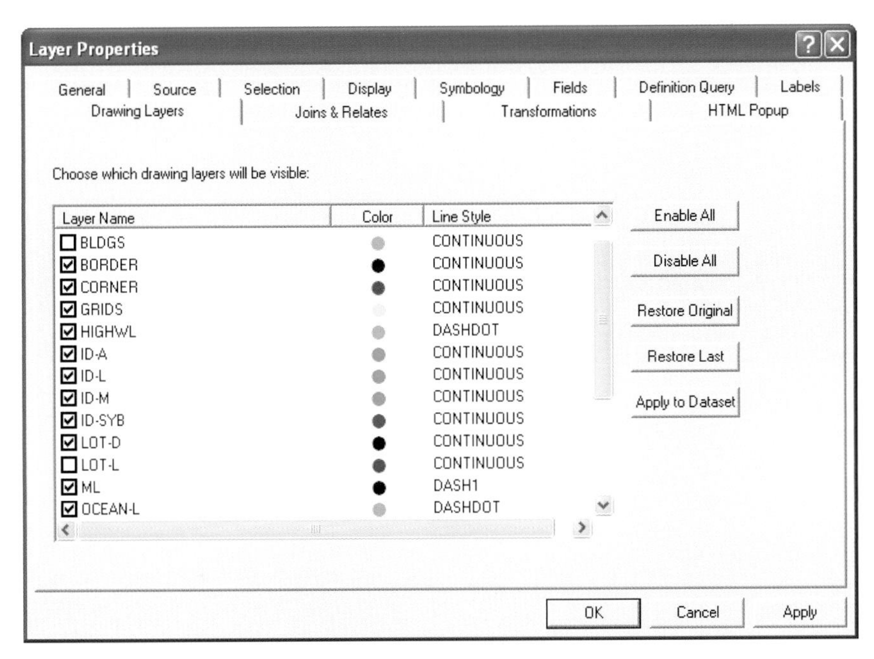

Figure 4–8 Turning off some layers for Parcels.dwg will clarify the relationship between the remaining polylines in Parcels.dwg and Index.dwg. (Note: not shown in the figure, but still turned off, is TREE-L.)

Roads are included in both Index.dwg and Parcels.dwg, but both sets of roads were drawn with the same blue color. Changing the draw symbol for both layers to different colors will make the relationship between the two layers easier to see (figure 4–9).

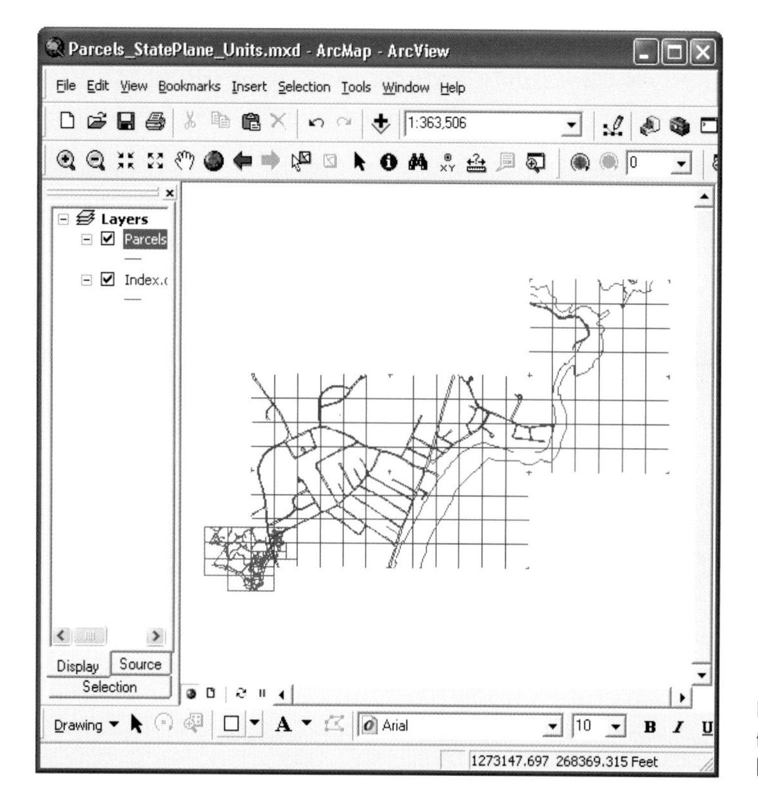

Figure 4–9 Changing the line color for the two layers makes it easier to distinguish between them.

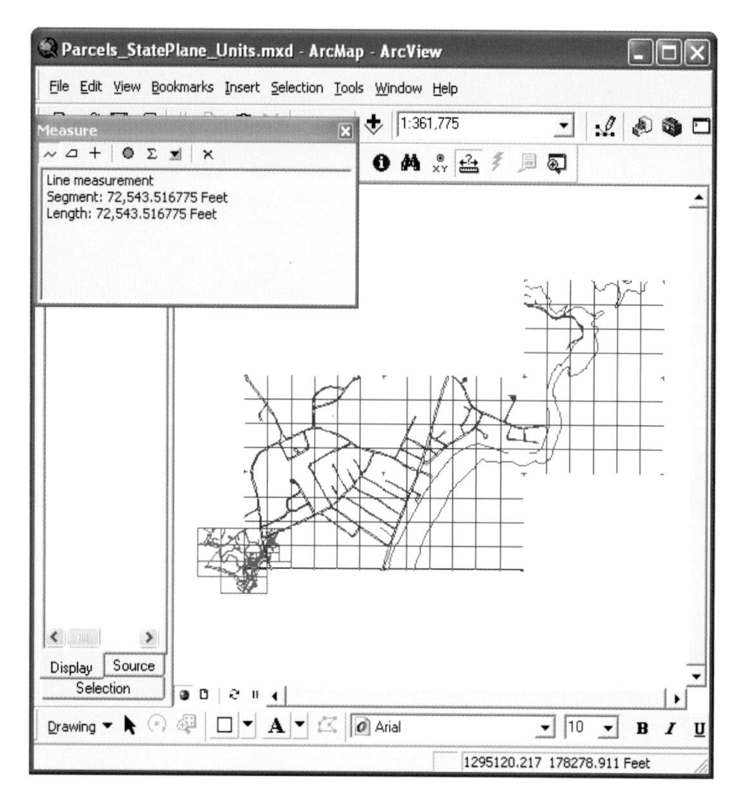

Figure 4–10 Measuring along the bottom edge of Parcels.dwg, the distance is more than 72,000 feet, or almost fourteen miles. That cannot be correct. You will need this measurement in order to compare with the length of the same line from Index.dwg, and use the two values to calculate units used when Parcels.dwg was created.

In order to find two common features that exist in both drawing files, we added the Annotation layers from both Parcels.dwg and Index.dwg to the ArcMap data frame, and opened the Annotation attribute tables (figure 4-11). Scroll to the right end of the tables to the TxtMemo field (field name is shown as "Text" in ArcMap 10). Scrolling through the TEXT values, you see street names; select the two annotations with the name CENTRAL from Parcels_inches.dwg, because that street name also exists in the TxtMemo field for Index.dwg.

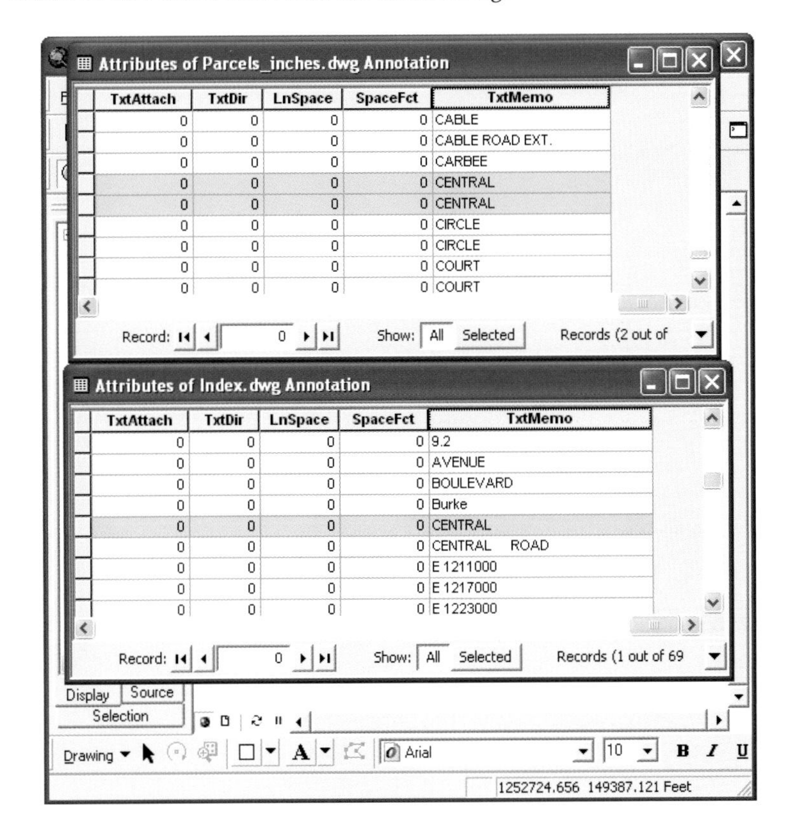

Figure 4-11 Selection of the street named CENTRAL from both annotation attribute tables. Note that in the new version of ArcMap, the field containing the text string will be named "Text".

Now zoom to Selected Features in the ArcMap data frame (figure 4-11). Since there are two roads named CENTRAL in Parcels.dwg, you have to examine the shape of these two roads, and compare them with the shape of the road named CENTRAL from Index.dwg to determine which is the matching feature.

Making that comparison, you can determine the location of the road named CENTRAL that exists in both drawing files. Measuring the length of that road feature in both files, you find that the feature in Parcels.dwg measures 6,000 feet, while the same feature in Index.dwg measures about 500 feet. Since 6,000 / 500 = 12, the units used to create Parcels.dwg are inches.

Notice in these screenshots that the full name of Parcels.dwg is Parcels_inches.dwg. While we saved the file with this name in AutoCAD, be aware that the providers of your CAD files will not be this obvious.

Now that we have identified the units used to create the Parcels drawing file, we need to modify the projection file with units of inches, so Parcels and Index will line up in ArcMap.

- Go to View > Data Frame Properties > Coordinate System tab, and click Modify.

- Change the name of the projection file to a useful name. Remember to use underscores instead of spaces, and do not add any special characters to the file name.

- Change the units of the projection to Inch_US.

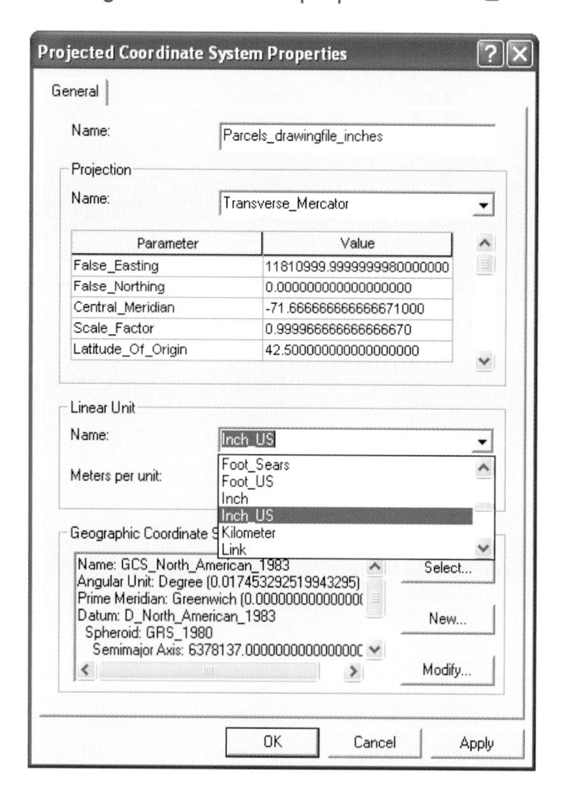

Figure 4–12 Custom projection file with a new name and corrected units. The false easting value has been recalculated automatically from survey feet to units of inches.

- Click Apply, OK, then click Apply again on the Data Frame Properties dialog box.

- Click Add to Favorites, then click OK again. Now the false easting and false northing will need to be adjusted to align Parcels.dwg with Index.dwg. (The results of these operations appear in figure 4–13.)

- Using the Measure tool, set the distance units to Inches, then measure the east-west distance in a straight line from the location of Parcels.dwg to a point directly south of Index.dwg. Go to View > Data Frame Properties > Coordinate System tab > Modify.

False easting and false northing adjustments

False easting and false northing values are sometimes inserted into a projection file in order to make all the x- and y-coordinate values across the area of the data positive numbers. False easting and false northing values can also be used to adjust the position of the data in either the east-west or north-south direction in order to align the data.

Making the false easting value in the projection file larger will adjust the position of the data to the west, moving the data to the left in the ArcMap display. Making the false easting value in the projection file smaller will adjust the position of the data east, to the right in the ArcMap display.

Adjustments to the false northing value in the projection file will move the data display north or south, though they are not as intuitive as the false easting adjustments. Making the false northing value in the projection file *larger* will move the data display *south* in the ArcMap window. Making the false northing value *smaller* will move the data display *north* in the ArcMap window.

Keep these adjustments in mind while creating the custom projection file to align your data in ArcMap.

Parcels.dwg needs to be moved east, so the measurement must be subtracted from the current false easting value to make the false easting smaller. Enter the new false easting value, then click Apply and OK, then Apply, Add to Favorites, and OK on the Data Frame Properties dialog box.

After the adjustment to the false easting, Parcels.dwg needs to be moved north. Measure the distance from Parcels.dwg north to the approximate location of Index.dwg. Remember that in moving the data north, the false northing gets *smaller*. Since the false northing in this case is 0, the measurement will be entered as a negative number.

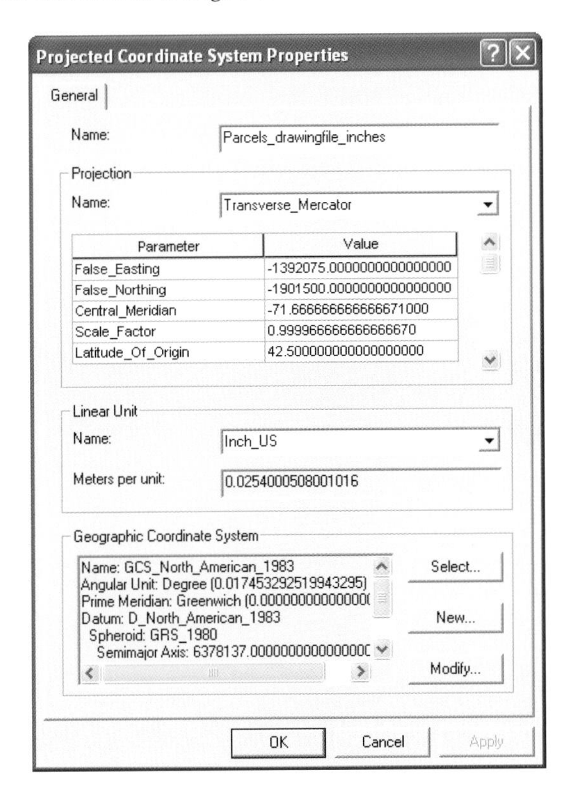

Figure 4-13 After the initial adjustments to the false easting and false northing, here are the values in the projection file.

Figure 4–14 shows the results of the initial adjustments. Parcels.dwg is much nearer to Index.dwg, but further adjustments are needed. Repeat measurements of the offset east-to-west to adjust the false easting and north-to-south to adjust the false northing, until the best possible alignment has been achieved between the two datasets.

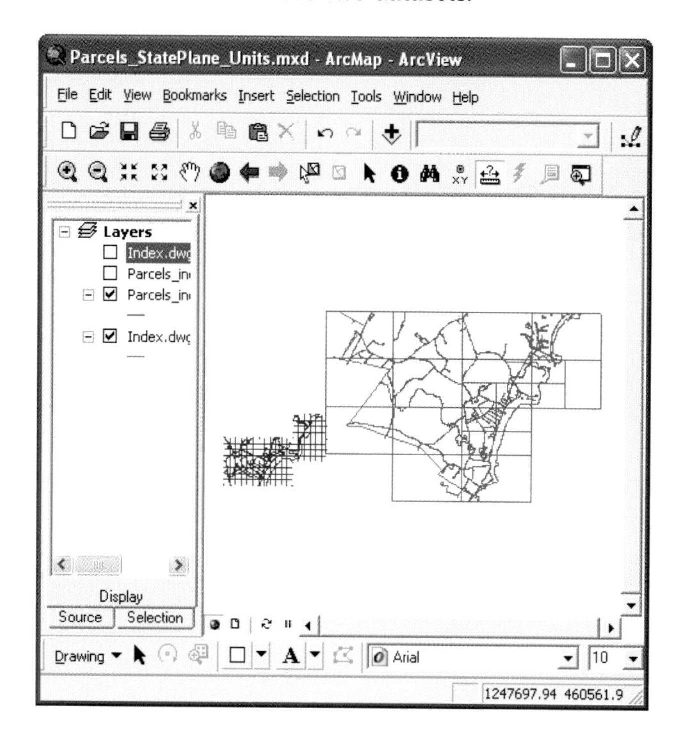

Figure 4–14 Here is the alignment of the two files after the initial adjustments. (The final alignment is shown in figure 4–16.)

After additional adjustments to the false easting and false northing, to move Parcels.dwg farther east and north, the final copy of the projection file is displayed in figure 4–15.

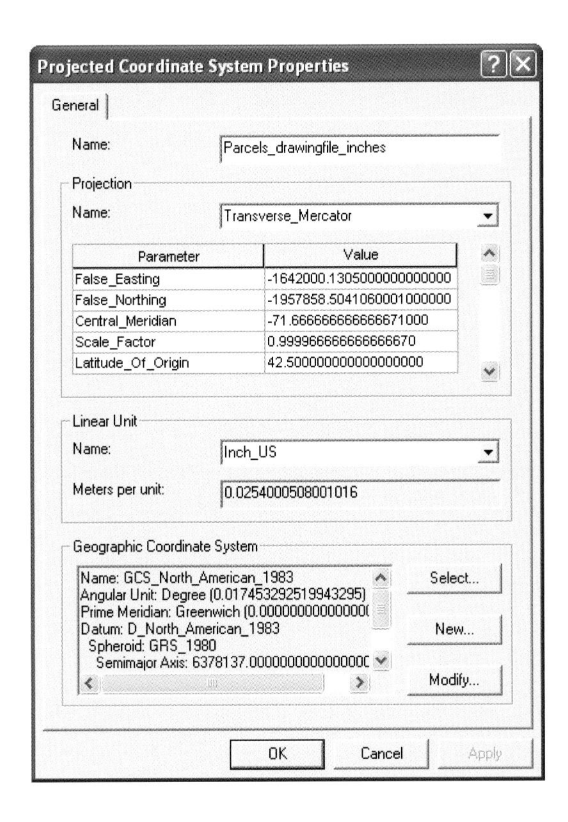

Figure 4-15 The final copy of the modified projection file, with units of Inch_US.

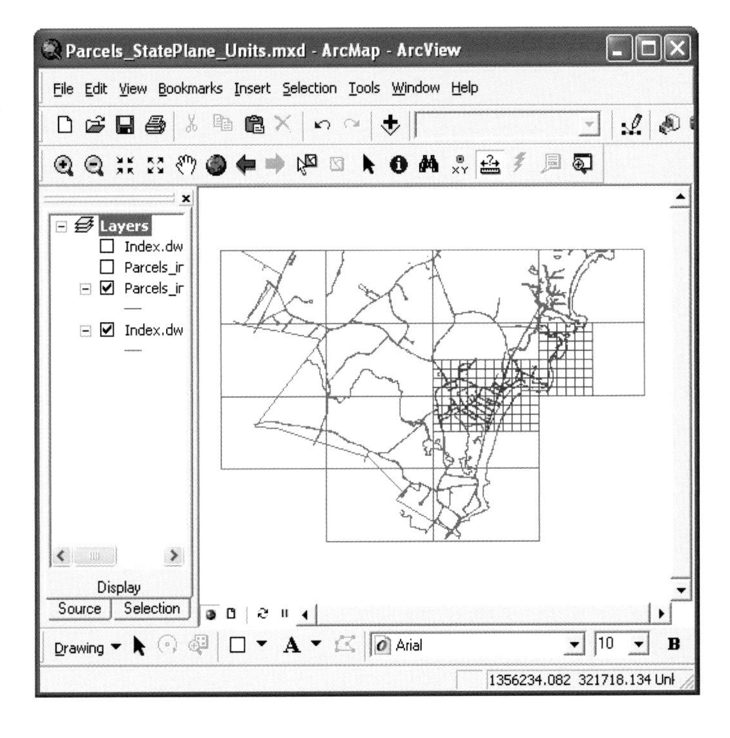

Figure 4-16 Final alignment of data in ArcMap.

Now that the data is finally aligned in ArcMap, you will need to save the custom projection file to disk for future use (page 59) or to define the projection using the custom projection file (page 60). Further instructions on saving and utilizing the custom projection file in ArcGIS Desktop follow the next section.

In this example, we customized a state plane coordinate system projection file using units of inches, in order to align an AutoCAD drawing file created in inches with Index.dwg that was drawn with units of U.S. survey feet (Foot_US). Next, we will examine the difference between units of the U.S. survey foot (Foot_US) and the international foot (Foot), and demonstrate the effect of the difference in these two units on the alignment of data in UTM and in state plane.

UNITS OF INTERNATIONAL FEET VS. U.S. SURVEY FEET

If there is a small (∼3.2 foot or 1 meter), consistent east-west shift across the entire extent of the data, in the same direction, the data may have been created in UTM with units of international feet rather than U.S. survey feet.

To change the units from Foot_US to the international foot, go back to View > Data Frame Properties > Coordinate System tab > Modify. If you have named the current custom projection file to include _Foot_US, change the name again. You can simply add _IntlFt to the end of the name.

Click the Units drop-down list, then scroll up and select the unit Foot. This is the name used in ArcGIS for the international foot, and the "Meters per unit" box will display the value 0.3048.

Click Apply and OK, then Apply and OK again on the Data Frame Properties dialog box. If the small shift you noticed before is gone, the data was created in the UTM coordinate system, but with units of international feet.

Repeat the process of clicking Add to Favorites to write the edited projection file to disk.

Use of international feet versus U.S. survey feet will also cause data alignment issues in the state plane coordinate system. The coordinate systems for SPCS include false eastings and in some cases false northings. These values make all the x- and y-coordinates for data within the zone positive numbers. While the difference in length between the international foot and the U.S. survey foot is literally microscopic (0.0000006096 feet; 0.000007315 inches; or 0.000001858 millimeters.), when this difference is multiplied by the large values typical for the SPC system false easting, the offset can be dozens of feet or more. Here is an example:

False easting in U.S. survey feet for NAD 1983 Montana zone 2500 = 1968500

False easting in international feet for NAD 1983 Montana zone 2500 = 1968503.937007874

The difference for Montana, depending on the units used in the SPC system for this zone on the NAD 1983 datum, is nearly 4 feet.

An extreme example is the state of Michigan. This state's three FIPS zones have very large false easting values, so the difference between international and U.S. survey feet for the false easting values is twenty-six feet or more.

If you see inconsistent offsets between the reference data and the unknown data, this may be a data-quality issue that you will have to resolve by communicating with the data source, field checking, additional surveying, obtaining additional reference data, or other techniques to reconcile the differences.

SAVING THE CUSTOM PROJECTION FILE TO DISK FOR FUTURE USE

The customized projection file exists *only* in the ArcMap document at this time. Here are the steps to save the custom projection file to disk, so that you can use the projection file to define the coordinate system for the data in ArcToolbox. You will also be able to project other data to the customized coordinate system after saving the projection file (PRJ).

- Return to View > Data Frame Properties > Coordinate System tab and click Add to Favorites on the lower right of the dialog; then click OK.

- When you click Add to Favorites, the custom projection file is saved to a location on the hard drive. The default location where the file is saved will depend on the version of ArcGIS Desktop you are using and the operating system of your computer. Refer to appendix C for the specifics.

- Open Windows Explorer and navigate to the folder where the custom projection file has been saved. Right-click the custom projection file and select Copy.

- Navigate to the installation location for ArcGIS Desktop on your computer (refer to appendix B for the specific location) and open the Coordinate Systems folder.

- Make a new folder in the coordinate systems directory and paste the custom projection file into the folder. You might name the new folder "Custom PRJ Files" on your computer; any name can be used. You can also create additional folders in this directory, if you find it more convenient to create new folders to contain custom projection files created for particular projects.

Installation location

The default installation location for ArcGIS Desktop is listed in the table in appendix B. If you do not find the software installed at that path, you can perform a search to locate the installation on your computer: search for a folder named "Coordinate Systems."

If the Application Data folder is not visible in Windows Explorer, go to View > Folder Options > View tab, and change the option button to "Show hidden files and folders." Also uncheck the two boxes immediately below that button. Figure 4-17 illustrates this dialog. When Windows produces a warning, click Yes on the warning. At the top of the View tab, click "Apply to All Folders", then click Apply, and click OK. The Application Data folder will now be visible.

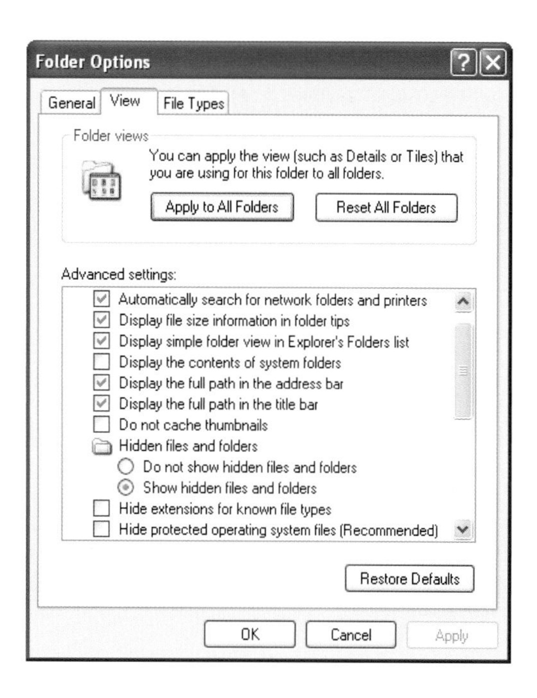

Figure 4-17 Folder Options dialog box in Windows Explorer, accessed to change display properties and make the Application Data folder visible.

DEFINING THE PROJECTION USING THE CUSTOM PROJECTION FILE

- Open ArcToolbox > Data Management Tools > Projections and Transformations > Define Projection. (To access this path in ArcGIS Desktop 10, from within ArcMap select Catalog window > Toolboxes > System Toolboxes.)

- From the "Input Dataset or Feature Class" drop-down list, select the data that was created with the custom coordinate system.

- Click Browse for the Coordinate System > Select. You will notice that you now have a new folder, along with Geographic Coordinate Systems and Projected Coordinate Systems, with the name you applied when creating the folder.

- Open the new folder, select the custom projection file, click Add, Apply and OK, and OK again on the Define Projection tool. The custom projection definition will be applied to the data and the data will project on the fly and align with other data in ArcMap.

DEFINING THE PROJECTION FOR CAD DATA

Additional steps are required when using the Define Projection tool to define the projection for CAD data. Instead of using the Define Projection tool, you can also use Windows Explorer and copy the file from the original location into the directory where the CAD file is saved, then rename the projection file to match the CAD file exactly. The projection file name is case-sensitive, so if the CAD file is named Parcels020609.dwg, the projection file will be named Parcels020609.prj.

More details for this process are provided in chapter 5.

SUMMARY

Chapter 4 describes techniques for identifying nonstandard units of measure that are sometimes used to create data and for modifying an existing projection file to incorporate these different units. The issue addressed in this chapter most often occurs with data created in CAD programs. CAD data can be created using units of U.S. survey inches, international inches, centimeters, millimeters, miles, kilometers, or other units of length. Be aware that this issue can also occur with other data formats, such as shapefiles or geodatabase feature classes. The same process outlined in this chapter can be applied to these other data formats to create a custom projection file and align the data.

The next chapter offers specific instructions for identifying these other units of measure. Chapter 5 also provides formulas you can employ to identify the units used when creating the data.

CHAPTER 5

ALIGNING CAD DATA — MODIFYING A STANDARD COORDINATE SYSTEM

"A CAD file I received from a client draws way off from my other data in ArcMap. What should I do?"

Computer-aided design (CAD) files, created with AutoDesk software such as AutoCAD or with Bentley Microstation, are supported in ArcGIS Desktop. Because these are drawing programs, used for surveying and engineering applications rather than geographic information systems, the file structure and units can be very different from those used in a GIS.

Data in GIS formats, like shapefiles and geodatabase feature classes, contains only a single feature type in the data layer. The feature type can be annotation (geodatabase only), points, polylines, polygons, or more complex features constructed from the basic feature types. In contrast, within a CAD file, data is divided into separate Layers (AutoCAD) or Levels (Microstation), and each layer or level can include all these feature types.

When the CAD file is added to ArcMap, each type of feature is grouped together, so all text entities, regardless of the layer or level on which they were created in the CAD program, draw together in ArcMap as Annotation. The same is true of point, polyline, and polygon features. The Layer name or Level number from the source data is preserved in the CAD file attribute tables when the data is added to ArcMap.

To align CAD data in ArcMap you will often need to create a custom projection file for the data. In selecting reference data to use in this process, a reference data layer that includes street names in the attribute table is ideal. You may need the street names in order to identify the exact location where the CAD data should be in relation to your reference data.

CAD data guidelines

You should request that the CAD operators providing data for use in your GIS follow these guidelines:

In AutoCAD

Create all data elements in Model space. ArcMap does not draw features created in Layout space.

Add layout elements such as the Legend, Scale Bar, Neatline, and publication credits to Layout space, not Model space.

Do not add Viewport entities to the drawing in Model space. Only use Viewports in Layout space.

Explode the blocks.

In Microstation

The seed file used to create new data must not contain features. The seed file must include only draw properties for features to be created in the new DGN file.

In both CAD programs

Do not save the CAD file with spaces in the file name. Use underscores.

Remove rotation from the file before saving. If the file must be rotated, addition of a north arrow is helpful to the GIS user. For AutoCAD users, add the north arrow in Model space.

Inform the GIS user of the units of measure used in creating the file.

Remove References when the drawing is completed, before saving the file. References attached to the CAD file in the native application will not display in ArcMap.

ISSUES TO CONSIDER WHEN WORKING WITH CAD FILES IN ARCMAP

A number of issues that may exist in the CAD file can affect display and use of the data in ArcMap. Several of these issues may exist in a single CAD file, affecting the alignment of the data when the file is added to ArcMap. Later in this chapter, you will find instructions for dealing with each of these issues listed by its number from the list that immediately follows here. It is advisable that you become familiar with all these issues. By doing so, you will recognize exactly what is affecting display of a CAD file when the data is added to ArcMap.

1. CAD files created in a standard coordinate system nonetheless may be using unusual units: centimeters, millimeters, inches, miles, or other nonstandard units. Working with these units of measure requires modifying a projection file in ArcMap to apply the units in which the CAD data was created. This may also require identifying the units used, if that information is not provided by the data source. (See pages 66–69.)
2. CAD files can be created with References attached. References are attached to the CAD file in the native application to add background vector or raster data. However, attached References in the file can affect the Extent of the data that is shown when the CAD file is added to ArcMap. Ideally, the References will be removed in the native application before the data is brought into ArcMap. (See page 69.)
3. CAD files can be created in a local coordinate system, with an arbitrary origin, but using standard units: the U.S. survey foot (Foot_US), the international foot (Foot), or meters. In order to align these data in ArcMap, a projection file installed with ArcGIS Desktop can be modified in ArcMap. (See pages 70-73.)

4. CAD data is frequently rotated in the native application, so that the CAD file will fit better on the page when printing. Ideally the rotation will be removed in the native application before providing the data for use in ArcMap. If not, a custom projection file can be created to compensate for the rotation in the data.

   Note that in ArcMap, standard data display places north at the top of the page. If the CAD file is rotated, north will be oriented in a different direction. Creating a custom projection file to compensate for rotated CAD data is addressed on pages 73–74 and in the next chapter, "Aligning rotated CAD data."

5. Various features from CAD files may not draw by default in ArcMap. Layers/Levels may be turned off, locked, or frozen in the native application. Features may be created with unsupported entities. Features inserted in BLOCKS (AutoDesk) or CELLS (Microstation) may not draw when added to ArcMap. SPLINE entities do not draw in ArcMap through version 9.3.1 although these entities will draw in ArcMap at version 10. HATCH fill patterns from CAD programs do not display in ArcMap because hatch patterns are symbology, not a feature. (See page 74.)

   In AutoCAD, "exploding the blocks" extracts the inserted features so that those features will draw in ArcMap. Using the tools in ArcToolbox to convert the CAD data to a geodatabase will also extract those features during the conversion process. Features contained in CELLS in Microstation DGN files are also extracted with these tools. (Further treatment of this and other conversion issues, however, is beyond the scope of this book. For instructions on converting CAD data refer to the ArcGIS Desktop Help installed with the software, or to the ArcGIS online documentation.)

Displaying AutoCAD MTEXT in ArcMap

Text in AutoCAD is sometimes created as MTEXT or multitext. This entity is a text box that can contain one or several lines of text. In the AutoCAD file, the MTEXT is inserted as a single box or block. When the annotation from an AutoCAD file containing MTEXT is drawn in ArcMap, the MTEXT will be drawn at the insertion point of the box, not at the position of the text as shown in AutoCAD. To display the text in the correct location in ArcMap, select the features in AutoCAD and explode the blocks.

6. In certain parts of the United States with higher elevations, such as the Rocky Mountain states, it is common practice to create CAD drawings in "ground" coordinates. If the CAD file is drawn in ground coordinates, aligning the data requires a "ground-to-grid" adjustment in the projection file when the data is added to ArcMap, in order to align with other data on a standard datum: NAD 1983, NAD 1983 (HARN), NAD 1927, or WGS 1984. This adjustment is handled by inserting or modifying a Scale_Factor parameter in a custom projection file, so that ArcMap can adjust the size of the data to match other existing data in grid coordinates. (See pages 74–75.)

7. Projections cannot be defined directly for CAD data that have spaces in the file name. Underscores or hyphens can be used in the file name instead of spaces. DWG and DGN files *cannot* be renamed in Windows Explorer, because these files are binary, and have the name embedded in the file contents. CAD data in these formats must be saved with a new name in the native application. In contrast, files with the DXF extension are text files, and these can be renamed using Windows rename function. The file extension .DXF must be retained.

   If you do not have AutoCAD or Microstation available, and the data source is not able to rename the file for you, the projection file may be named esri_CAD.prj instead. This will define the projection for all CAD files in the folder. Refer to page 73 or 78 for further discussion of this topic.

PROCEDURES TO ADDRESS THESE ISSUES

The following instructions apply to the numbered items in the introductory section above. As mentioned earlier, though, several of these issues may exist in a single CAD file. The interaction between some of them can affect alignment of the data.

1. IDENTIFYING UNITS OF MEASURE

As discussed in the preceding chapters, AutoCAD DWG/DXF files and Microstation DGN files can be created in a projected coordinate system such as state plane, but nonstandard units are often used when the file is created. These units can include centimeters, millimeters, inches, kilometers, miles, and units of a tenth of an inch or 1/1,000th of a foot.

If the CAD data source provides the coordinate system information, including the units used to create the file, you can create a custom projection file in ArcMap using the specified parameters and units. More commonly, though, the CAD data source will only provide the file, leaving you to decipher the coordinate system for the data and figure out what units were used to create the data. If you have access to the native application (AutoCAD or Microstation), you can easily read in that application the units used to create the CAD data. If the original CAD program is not available, however, you will have to identify the units in ArcMap by comparison with reference data and careful measurement. You can start the process this way:

Microstation uses units of international feet, rather than U.S. survey feet by default. If the DGN file is created with units of feet, check the units carefully to be sure the correct definition for the foot is used when modifying or creating a custom projection file.

To identify the units of measure, open ArcMap with a new, empty map. Navigate to the folder on the local hard drive where the CAD file is saved, double-click the CAD file icon, and add only the Annotation and Polyline feature classes from the CAD file to ArcMap.

When the "Missing spatial reference" warning appears, click OK to add the data to ArcMap. Add your reference data to the ArcMap document. The reference data must be in a projected coordinate system, with units of feet or meters, because CAD data is never created in geographic coordinates with units of decimal degrees.

Right-click the name of the Polyline layer from the CAD file > Properties > Source tab.

In the Extent box at the top of the tab, count the number of digits to the left of the decimal for the Top, Bottom, Left, and Right coordinates. If the CAD file was created with units of inches, centimeters, millimeters, tenths of an inch, or other *small* nonstandard units, the number of digits to the left of the decimal can be as many as 14, instead of 6, 7, or 8 as would be expected with units of feet or meters for data in a standard coordinate system.

Here is a list of some techniques you can use to identify the units used in creating the CAD file:

A. Is a scale bar included in the CAD drawing? The scale in CAD drawings will most often be represented as something like 1" = 400' (1 inch on the page = 400 feet on the ground) or 1" = 100 meters (1 inch on the page = 100 meters on the ground). A scale bar—a line of a specific length and distance—may also be included.

The units to the right of the equal sign in the scale as expressed above, or the units of measure indicated on the scale bar, will normally be the units used to create the file in the CAD program.

B. Set the units for the ArcMap data frame to the units indicated in the CAD file scale bar to start with, so the Measure tool will work. Since the first data added to the ArcMap data frame (the CAD data) does not have a coordinate system defined, ArcMap does not currently have units of measure set for the data frame, so the Measure tool is out of commission.

To set the units, go to View > Data Frame Properties > General tab (see figure 5-1), and set both the Map and Display units to match the units from the CAD file scale bar, then click Apply and OK.

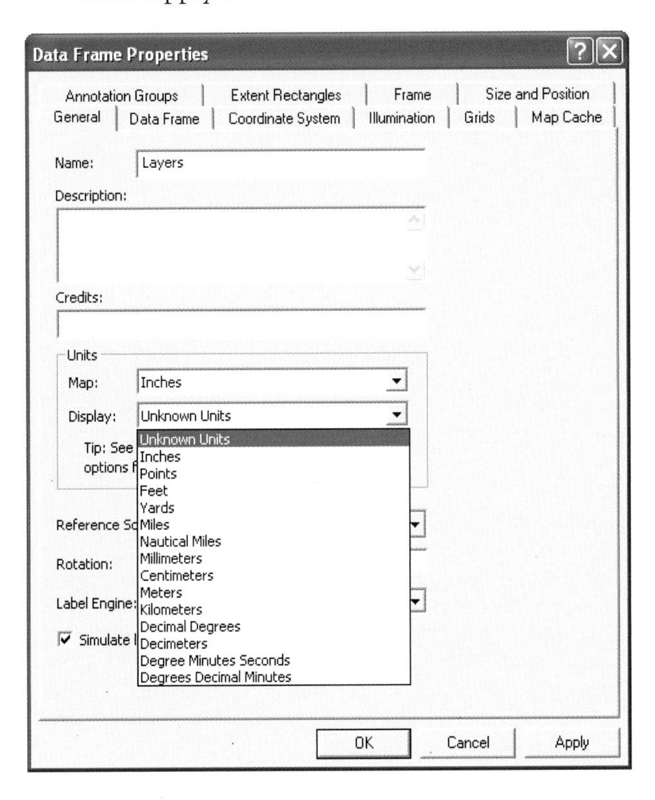

Figure 5-1 Data Frame Properties > General tab, setting map and display units for the ArcMap data frame. Both Map and Display must be set to the same units of measure.

Click on the Measure tool, set the Distance units to match the units set for the ArcMap data frame, zoom in to the scale bar in the CAD file, set Snapping in the Measure tool, then measure the length of the scale bar.

Does the Measure tool output match the displayed length of the scale bar? For example, if the scale as shown says 1" = 400', 1" divisions on the scale bar should each measure 400 feet. If the scale reads 1" = 100 meters, the Measure tool should return 100 meters as the length of a division in the scale bar. If the scale bar returns the correct measurement, you know that the units applied to the ArcMap data frame are those used to create the CAD file.

If the Measure tool returns some other value or if no scale bar is provided in the CAD file, continue the C and D instructions that follow here.

C. Open the attribute table for the CAD Annotation layer, and scroll across to the TxtMemo field at the right-hand end of the attribute table. This field lists the text strings that have been added to the CAD file. Carefully scroll down through the listed text strings and look for values that appear to be measurements. CAD files very often contain parcel dimensions such as 122.3'—or dimensions for the length of a water or sewer pipe—that you can locate and measure in the ArcMap data frame. When you find a field populated with a value that appears to be a measurement, select that record in the table. Click on Selection > Zoom to Selected Features in the top bar of the ArcMap window. The selected text will be centered in the ArcMap data frame.

Set the Map and Display units for the ArcMap data frame as described above, setting the units to match the units from the dimension text. In the Measure tool, set the Distance units, also matching the units from the dimension text, turn on Snapping, then measure the line labeled with the selected text.

Using the example above, if the dimension for the line is marked as 122.3' and the measurement in ArcMap returns 122.3', you know the units used to create the CAD file are feet. If a different number is returned by the Measure tool, perform the units conversion as shown in these sample calculations:

If you get 1,467.6 feet, the units used to create the file are inches.

122.3 feet × 12 inches/foot = 1,467.6 inches or

1467.6 ÷ 12 = 122.3 feet

If the Measure tool returns a number such as 37 feet, the units used to create the file are meters.

122.3 feet × 1200/3937 = 37.27712... meters or

37.27712 × 3937/1200 = 122.3 feet

If the Measure tool returns a number like 3,727.712... feet, the units used to create the file are centimeters.

122.3 feet × 1200/3937 × 100 = 3,727.712... centimeters or

3727.712... × 3937/1200 ÷ 100 = 122.3 feet

If the Measure tool returns something like 37,277 feet, the units are millimeters.

122.3 feet × 1200/3937 × 1000 = 37,277.12 millimeters or

37277.12... × 3937/1200 ÷ 1000 = 122.3 feet

These formulas can be applied to any measurements found in the file, in order to determine the units used when creating the CAD file.

D. If no scale or scale bar exists in the CAD file and no dimension text exists, go through the steps under heading number 3 on the next page to begin creating a custom projection file for the CAD data. Aligning the CAD data in the same general location as the reference data will allow you to more easily identify and measure the same feature in the two sets of data side by side. You can compare the length of a street segment or other line feature in the CAD file with a known distance from the reference data to calculate the units used to create the CAD file. (Chapter 4 describes this process for an AutoCAD drawing file created with units of inches.)

2. CAD FILES CREATED WITH REFERENCES ATTACHED

In ArcMap, you can add to the map document many datasets containing annotation, points, polylines, polygons, as well as raster images. With their coordinate systems defined correctly, the data will align, and highly detailed maps can be created. In CAD programs, however, only a single file can be opened at one time in the application. To display other files or images as background for the open file in the CAD program, additional vector or raster datasets are attached to the CAD file as References.

For all types of CAD data, References attached to the file are "anchored" to the file, and the "anchor" points of the References are often at 0,0, the lower left corner of the CAD file. When examining the extent of the CAD feature class in ArcMap, this results in Left and Bottom coordinates that are zeros or other very small numeric values, even though the file may have been created in standard state plane or UTM coordinates.

A second issue also exists with Microstation DGN files that have references attached. When added to ArcMap, in the Layer Properties > Drawing Layers tab, *all* levels that exist in both the original DGN file and the referenced files will be listed. For example, Level 0 always exists in a DGN file, even if that Level is empty and contains no features. If the DGN file has nine references attached, Level 0 will be listed ten times on the Drawing Layers tab. The first "Level 0" listed is the one associated with the current DGN file: the other nine are from the Reference files. This applies to all Levels for the DGN file, and a DGN file can include hundreds of Levels. No data will be drawn in ArcMap from the referenced DGN files because the data from the References does not exist in the current DGN, but this complicates the Level listing on the Drawing Layers tab.

The best practice is to avoid these issues by removing all references attached to CAD files in the native application and resaving the file, before you bring the CAD file into ArcMap. If you have the original application available, remove the references from the file. If the CAD file is provided by another party, request that references be removed before the file is sent to you.

If references are not and cannot be removed from the file in the native application, you will need to rely solely on the Top and Right coordinates to determine the coordinate system for the data.

3. IF CAD DATA IS IN NONROTATED LOCAL COORDINATES

To a new, empty ArcMap session, you have added the polyline and annotation layers from the CAD file together with the reference data the CAD file is supposed to line up with. The reference data must be in a projected coordinate system with units of feet or meters, and must have the coordinate system defined. It is preferred that the units of the coordinate system for the reference data match the units used to draw the CAD file in the native application. It is strongly recommended that you use reference data that includes streets and street names in the attribute table, if such data are available.

Temporarily uncheck the Annotation layer in the ArcMap Table of Contents so it does not draw.

Reference data resources

Street data with street names as attributes is available from many sources. StreetMap data is provided with ArcGIS Desktop on the Data & Maps DVDs/CDs. Use of the detailed StreetMap data requires an additional license, but a sixty-day evaluation license is available through your ESRI sales representative or through ESRI Customer Service.

NAIP imagery is available for download from the USDA's Geospatial Gateway at

`http://datagateway.nrcs.usda.gov/`

ArcGIS Online is also a resource for imagery:

`http://resources.esri.com/arcgisonlineservices/index.cfm?fa=home`

If the CAD file contains points that represent survey control points (benchmarks) placed by the U.S. Geological Survey, a shapefile with the locations of these benchmarks for the quad or county in which the data is located, in GCS_North_American_1983_HARN, can be downloaded at

`http://www.ngs.noaa.gov/cgi-bin/datasheet.prl`

You can identify the 1:24,000 quad in which the data is located by adding usgs24q.shp to ArcMap with your data. This shapefile is installed with ArcMap in the Reference Systems folder, or another install location. (Refer to appendix B for the default installation location.) The attribute table for the shapefile contains quad names and their USGS ID numbers.

TIGER/Line data by county including streets and street names, is available from the U.S. Bureau of the Census or can be downloaded in shapefile format, free of charge, from the ESRI Web site at:

`http://arcdata.esri.com/data/tiger2000/tiger_download.cfm`

This data is in geographic coordinates, on the NAD 1983 datum, and must have the coordinate system defined before using the data as a reference layer in ArcMap. If employing this data as reference, first use the Define Projection tool in ArcToolbox, then the Project tool to project the data to state plane or UTM, so that the reference data will have linear units of measure.

The accuracy of TIGER data varies substantially in different counties within the United States. It is very useful for providing street names, but other resources should also be used to obtain more precise data alignment.

Go to View > Data Frame Properties > Coordinate System tab. In the lower window labeled "Select a coordinate system," open the Layers folder, and open the folder labeled with the name of your reference data file. Click on the coordinate system definition name for that data. This is the base coordinate system you will modify to create the custom coordinate system for your CAD data. Click Apply and OK to set the coordinate system for the ArcMap data frame.

Click the full extent button. The CAD data will display southwest of the reference data, *if the units used to create the file were standard feet or meters*. If the CAD file was created with units of inches, centimeters, millimeters, or other small units, the CAD polylines will draw northeast of the reference data.

If the CAD polylines are not visible on the screen at the full extent, right-click the name of the CAD polyline layer and select Zoom to Layer. On the Drawing toolbar at the bottom of the ArcMap window, click the drop-down list next to the New Rectangle tool (the white square) and select New Marker from the bottom row. Drop a marker symbol in the center of the CAD data then zoom to the full extent again. You still won't be able to see the CAD data, but the marker symbol will show the location of the CAD polylines in relation to your reference data.

Click the Measure tool and set the Distance units to match the units for the coordinate system of your reference data (feet or meters). Measure the approximate distance between the CAD polylines and the data in the real-world coordinate system, in the east-west direction only. Do not measure on the diagonal, just a straight line west to east (left to right). At this point, rounding measurements to the nearest one hundreds is sufficient, and there is no need to keep track of decimal places. The data is offset too far for accurate measurements to be collected.

It helps to write down these measurements on a piece of paper with a directional arrow indicating the direction the CAD data needs to move to align with the reference data.

Go to View > Data Frame Properties > Coordinate System tab, and click on the Modify button.

Enter a new name in the top box for the custom projection file. Any name can be used as long as it does not contain spaces or special characters; underscores are acceptable.

Under Parameters, there is a Value listed for the false easting. Making the false easting value larger moves the CAD file west. Making it smaller moves the CAD data east.

Assuming that the CAD data is drawn west of the reference data, you will subtract the measured distance from the existing false easting value, and enter the new value in the field. This new value may be either a positive or negative number. Do not include commas as separators in the number. If the CAD data is displayed east of the reference data, you will add the east-west measurement value to the false easting.

ArcMap calculates position to sixteen significant digits from the projection file, so zeroes to the right of the decimal point should be saved in these Value fields.

Note that at this stage of this process, there is no need to keep track of decimal places in these values. Numbers to the right of the decimal will become important when the data is aligned more closely, but rounding measurements to the nearest hundred at this stage is sufficient.

Click Apply and OK on the Projected Coordinate System Properties dialog box. Click Apply on the Data Frame Properties dialog box.

The CAD data will move east (or west) to more closely align with the reference data. After the false easting adjustment, the CAD polylines should approximately align with the reference data in the east-west direction. The CAD polylines will still be offset in the north-to-south direction.

Click Add to Favorites. That writes a copy of the custom projection file to disk. Click OK on the Data Frame Properties dialog box.

Delete the marker symbol that showed the previous location of the CAD file, then zoom to the full extent again.

Use the Measure tool again, but this time measure the offset between the CAD data and the reference data in a direct line north to south. It can help to make a note for yourself: write down this distance value, with an arrow indicating the direction the CAD data is supposed to move in relation to the reference data.

Go back to View > Data Frame Properties > Coordinate System tab > Modify button.

The north-south adjustment is made by changing the false northing value.

The adjustment to the false easting seems logical. When you make the false easting number larger, the CAD data moves west, to the left on the screen. When you make the false easting number smaller, the CAD data moves east, to the right on the screen. But the adjustment of the false northing value may seem backwards to you. Making the false northing value *larger* moves the CAD data *south*. Making the false northing *smaller* moves the data *north*. Regardless, this is the way it works.

Assuming that the CAD polylines are now *south* of the reference data, *subtract* the measured distance from the existing false northing value. If the false northing is 0, enter the measured north-south offset as a negative number, remembering to keep the zeroes to the right of the decimal.

If the CAD polylines are *north* of the reference data, you will *add* the measured distance to the existing false northing value. If the false northing is 0 you will enter the north-south offset as a positive number.

Click OK, Apply, Add to Favorites, and OK.

Repeat these steps as needed, making incremental adjustments to the false easting and false northing, until the alignment of the CAD polylines to the reference data is the best it can be. Keep in mind, the CAD data may not align exactly with your reference data because of inaccuracies in one or both sets of data.

When you have made sufficient adjustments to the projection file so that the CAD polylines are drawing in the correct city or county area, but you cannot identify the exact location where the CAD data should be in relation to your reference data, use the CAD Annotation layer, together with attributes from the reference data, to align the data more precisely.

Check the CAD Annotation layer box in the table of contents to draw the Annotation. To make sure all Layers/Levels for the CAD Annotation are turned on, right-click the name of the CAD Annotation layer, select Properties > Drawing Layers tab, check Enable All, then click Apply and OK. Right-click the CAD Annotation layer name again, and open the attribute table. Scroll to the right end of the attribute table, to the TxtMemo field, and examine the text strings, looking for a street name. When you find a street name, select that record in the table.

Open the attribute table for your reference data, and locate the same street name. Select that record.

In the top bar of the ArcMap window, click Selection > Zoom to Selected Features. You will be able to see the locations of the selected street name in the CAD Annotation layer and the selected street from your reference data. Make additional adjustments to the false easting and false northing values in your custom projection file to line up the streets from the CAD file, and the reference data, in the ArcMap window.

When the final version of the projection file has been achieved, click Add to Favorites one more time. Click Apply, then click OK in the Data Frame Properties dialog box. You may find it useful to save the MXD file.

When you click Add to Favorites in the Data Frame Properties dialog box, the custom projection file is saved to a specific location on your computer. The path to that location will depend on the version of ArcGIS Desktop you are using as well as the operating system of your computer. Refer to appendix C to find this specific location.

Open Windows Explorer (Right-click Start > Explore) and navigate to the path where the custom projection file has been saved. If folders in the path are not visible, go to View > Folder Options > View tab, and change the option button to "Show hidden files and folders." Also uncheck the two boxes immediately below that button in the dialog. When Windows produces a warning, click Yes on the warning. At the top of the View tab, click "Apply to All Folders", then click Apply and click OK. The hidden folders are now visible.

Right-click the custom projection file, select COPY, then navigate to the install location for your version of ArcGIS Desktop (see appendix B for the exact default location).

Make a new folder in the Coordinate Systems directory and paste the custom projection file into the new folder. "Custom PRJ Files" is one folder name you could use, but any name will do.

Still in Windows Explorer, navigate to the folder where the CAD file is located on the local hard drive. Paste the custom projection file into that folder, and rename the PRJ file to match the name of the CAD file exactly. This is case-sensitive. For example, if the CAD file name is Parcels022007-G.dwg, name the projection file Parcels022007-G.prj.

If multiple CAD files in the same local coordinate system are stored in the folder, the custom projection file can be named "esri_CAD.prj". ArcMap will recognize this projection definition and will apply the custom projection to all CAD files in the directory when the files are added to ArcMap. When the projection has been defined for the CAD files, the data will project on the fly in ArcMap and line up with your other data in any new map documents you create.

The projection file named esri_CAD.prj will define the projection for CAD files with spaces in the name.

4. ROTATED CAD FILES

When you see a CAD drawing depicting features in an area where the greatest extent runs diagonally southwest to northeast or northwest to southeast, the data may have been rotated in the CAD program, to fit more neatly on the page when printing. If so, a north arrow will usually be included in the drawing to indicate the direction of true north.

If you receive a CAD file in which the data is rotated, the data will also appear rotated when the data is drawn in ArcMap. If at all possible, have the data source remove the rotation from the file, so that north is at the top of the page and the geographic orientation of the data is correct when the data is added to ArcMap. Often, though, this is not possible.

Frequently the rotation in the CAD file will not be apparent until you have gone through the steps (outlined under issue number 3) to align the CAD data (more or less) with the reference data in ArcMap. When the CAD data is drawn in the same area as the reference data, you may suddenly realize that the CAD polylines are at the wrong angle in relation to the reference data.

The solution is to create a different custom projection file, using a projection that includes an azimuth and/or rotation parameter, in order to compensate for the rotation and align the data.

You can use the following projections, available in ArcGIS Desktop, to compensate for a rotation in the CAD data:
Local
Rectified skew orthomorphic center
Rectified skew orthomorphic natural origin

Turn to the next chapter for specific instructions on creating a custom projection file that includes a rotation and/or azimuth parameter, in order to adjust for rotation in the CAD file.

5. IMPROVING LAYER/LEVEL VISIBILITY

Unlike GIS data formats in which data is grouped into data layers by feature type, CAD data is divided into Layers (AutoDesk products) or Levels (Microstation). Each Layer or Level can contain Annotation, point, polyline, polygon, and multipatch features. When the CAD data is added to ArcMap, the data is divided by the type of feature so all polylines, regardless of the Layer or Level on which each feature was created in the CAD program, are grouped together in ArcMap.

Sometimes, to prevent the data from drawing or from being edited, Layers or Levels are turned off, frozen, or locked in the native CAD application. If so, the data on that Layer/Level will not draw by default when the data is added to ArcMap.

To ensure that all data from the file is drawn when adding polylines or other feature types from the CAD file to ArcMap, right click the name of the CAD data layer in the ArcMap Table of Contents > Properties > Drawing Layers tab. Any Layer/Level name that is unchecked in the list will not draw in ArcMap, and data from that layer will not be listed in the attribute table. Note, however, that empty Layers or Levels can exist in the CAD file; the Layer or Level may have been created in the CAD file, but the Layer or Level might not contain any features.

To activate all Layers/Levels for drawing in ArcMap, click Enable All, then click Apply and OK.

Recall that features from Levels in Microstation DGN files that are listed from references attached to the DGN will not display since those features do not exist in the DGN file currently displayed in ArcMap.

Of course Layers/Levels can also be turned off in the same dialog. If the CAD file contains too much data, it can be difficult to distinguish parcel lines (or other features required to align the data) from contour lines, water and sewer lines, vegetation, etc. Use the Identify tool in ArcMap to find out which Layers/Levels contain the data of interest, then turn off the Layers or Levels that are not relevant for the alignment process.

6. IF CAD FILES WERE CREATED IN GROUND COORDINATES

The CAD data source will nearly always provide the scale factor used when creating the file if the file was created in ground coordinates.

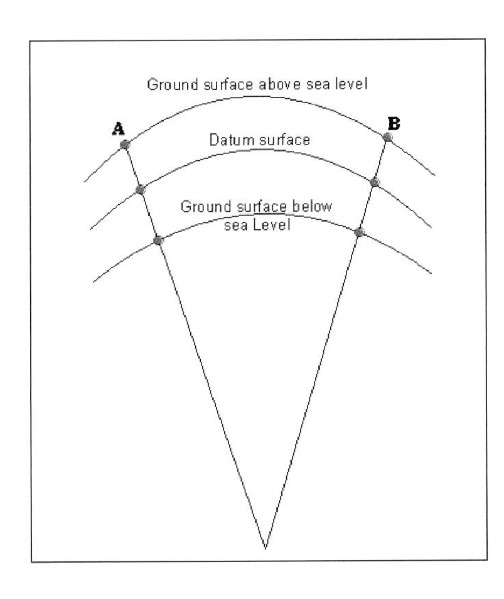

Figure 5–2 Illustration of the difference between ground and grid coordinates. The distances between points A and B on the ground surfaces will be longer or shorter than the same distance on the datum surface depending on whether the ground surface is above or below the surface of the datum.

In figure 5-2, compare the length—between points A and B—of the lines labeled "Ground surface above sea level," "Ground surface below sea level," and "Datum surface." The lines extend between the same two points, but the line on the ground surface is longer or shorter than the same line between those points at the level of the datum, depending on the elevation where the data were collected.

Let's say that the line on the ground has been surveyed very accurately and we know it to be 1,000 feet long. The scale factor used for this area, which is about 11,000 feet above sea level, is 1.0004. When the 1,000-foot line is dropped to the level of a datum, the length of the line on the map will be 999.6 feet (1,000 ÷ 1.0004) or 999 feet, 7.2 inches.

If the base projection is Lambert conformal conic, in order to align CAD data created in ground coordinates with other data projected on a datum (such as NAD 1983), this is what needs to be done: the scale factor parameter, with the value provided, must be inserted into the projection file used to define the coordinate system for the CAD data.

Let us assume that the CAD file is created using scaled coordinates, in the state plane coordinate system, for an area in which the *base* projection is Lambert conformal conic. Recall from chapter 3 that state plane zones that are broader east-west than north-south have coordinate systems based on Lambert conformal conic. For this example, we will use the projection file NAD 1983 StatePlane Colorado Central FIPS 0502 (Feet):

PROJCS["NAD_1983_StatePlane_Colorado_Central_FIPS_0502_Feet",GEOGCS["GCS_North_American_1983",DATUM["D_North_American_1983",SPHEROID["GRS_1980",6378137,298.257222101]],PRIMEM["Greenwich",0],UNIT["Degree",0.0174532925199432955]],PROJECTION["Lambert_Conformal_Conic"],PARAMETER["False_Easting",3000000.000316083],PARAMETER["False_Northing",999999.999996],PARAMETER["Central_Meridian",-105.5],PARAMETER["Standard_Parallel_1",38.45],PARAMETER["Standard_Parallel_2",39.75],PARAMETER["Latitude_Of_Origin",37.8333333333333334],UNIT["Foot_US",0.304800609601219241]]

The Scale_Factor parameter is inserted into the projection file as shown in the sample below. Note the location where the Scale_Factor parameter is inserted, and that the name of the projection file has been changed.

PROJCS["NAD_1983_SP_CO_Cen_FIPS_0502_Feet_scaled",GEOGCS["GCS_North_American_1983",DATUM["D_North_American_1983",SPHEROID["GRS_1980",6378137,298.257222101]],PRIMEM["Greenwich",0],UNIT["Degree",0.0174532925199432955]],PROJECTION["Lambert_Conformal_Conic"],PARAMETER["False_Easting",3000000.000316083],PARAMETER["False_Northing",999999.999996],PARAMETER["Central_Meridian",-105.5],PARAMETER["Standard_Parallel_1",38.45],PARAMETER["Standard_Parallel_2",39.75],PARAMETER["Scale_Factor",1.0004],PARAMETER["Latitude_Of_Origin",37.83333333333334],UNIT["Foot_US",0.304800609601219241]]

To modify the projection file, open Windows Explorer and navigate to the **Coordinate Systems \Projected Coordinate Systems\State Plane** folder, and open the appropriate folder to locate the state plane projection file, with the correct datum and units, for your area.

Open the file with NotePad. *Do not use WordPad.* WordPad adds formatting characters to the file, which will corrupt the file. If the file contents appear as a single long line, click on the Format button then check Word Wrap.

Insert the following string with the scale factor value provided by the data source into the projection file, between the parameters Standard_Parallel_2 and Latitude_of_Origin.

Note: if the data location is above sea level, the scale factor value must be greater than 1. If the data source provides a scale factor value that is less than 1 this number is the inverse of the scale factor. To calculate the true scale factor for insertion into the projection file, you will divide 1 by the number provided that is less than 1. The output will be the value to use.

PARAMETER["Scale_Factor",1.0004],

There are no spaces. The uppercase and lowercase letters and punctuation must be entered exactly as shown, including the trailing comma.

The projected coordinate system name in the top line in the example above must also be changed to indicate that the file has been modified. It's a good idea to do so by inserting the text "_scaled" at the end of the projection file name.

When these two changes have been made in NotePad, click File > Save As, and navigate to the Coordinate Systems directory in the ArcGIS Desktop install location. In that folder, create a new folder named Custom PRJ Files or whatever name is useful to you.

In the NotePad Save As dialog at the bottom, change Type to "All files". Encoding remains ANSI. Save the file in the Custom PRJ Files folder created above, with a name that includes the word "scaled" and the PRJ extension.

In ArcMap, open View > Data Frame Properties > Coordinate System tab. In the lower window labeled "Select a coordinate system", open the Predefined folder. You will now see the new folder named Custom PRJ Files (or the name you assigned). Open that folder, and select the scaled projection file, then click Apply and OK on the Data Frame Properties dialog box.

Is the CAD data now the same size as your reference data? If so, the scale factor is correct and you can use this projection file to define the coordinate system for the CAD data.

THE SCALE FACTOR MAY BE APPLIED TO THE FALSE EASTING AND FALSE NORTHING

In some cases, the scale factor not only is inserted into the projection file but also applied to the false easting and false northing values. This will make the false easting and false northing slightly larger than the standard values defined by the National Geodetic Survey. If this change has been made to the projection file, the CAD data will display west and possibly north of the reference data in ArcMap, and the offset will be consistent across the entire extent of the data.

Using the example above for "NAD 1983 StatePlane Colorado Central FIPS 0502 Feet Scaled", the standard values for the false easting and false northing are 3000000.000316083 and 999999.999996.

Multiply the false easting value times the scale factor value, and carefully enter the new false easting value into the projection file. In this case the math is

False easting 3000000.000316083 × 1.0004 = 3001200.0003162094332

False northing 999999.999996 × 1.0004 = 1000399.9999959984

Notice that applying the Scale_Factor to the false easting offsets the data by about 1,200 feet. The adjustment to the false northing offsets the data by about 400 feet.

This can be done in ArcMap at View > Data Frame Properties > Coordinate system tab > Modify.

Notice that the Scale_Factor parameter inserted into the projection file in NotePad now appears in the parameters listed in the Projected Coordinate System Properties dialog box.

Alter the name of the projection file again, adding something like "_fe" to indicate the false easting has been altered, then enter the new values for the false easting and false northing that have been calculated to incorporate the scale factor. Here is the sample custom projection file that now includes the modified false easting and false northing:

PROJCS["NAD_1983_SP_CO_Cen_FIPS_0502_Feet_scaled_fe",GEOGCS["GCS_North_American_1983",DATUM["D_North_American_1983",SPHEROID["GRS_1980",6378137,298.257222101]],PRIMEM["Greenwich",0],UNIT["Degree",0.0174532925199432955]],PROJECTION["Lambert_Conformal_Conic"],PARAMETER["False_Easting",3001200.0003162094332],PARAMETER["False_Northing",1000399.9999959984],PARAMETER["Central_Meridian",-105.5],PARAMETER["Standard_Parallel_1",38.45],PARAMETER["Standard_Parallel_2",39.75],PARAMETER["Scale_Factor",1.0004],PARAMETER["Latitude_Of_Origin",37.83333333333334],UNIT["Foot_US",0.304800609601219241]]

Click OK, then click Apply and OK on the Data Frame Properties dialog box.

If the base projection of the state plane PRJ file being modified is transverse Mercator, which already includes the Scale_Factor parameter, the new scale factor value is simply typed into the Value field in ArcMap > Data Frame Properties > Coordinate System tab > Modify dialog box. Remember that the projection file name must be changed in the dialog box.

IF THE DATA LIES BELOW SEA LEVEL

There are places in the world that are below sea level, although the only location in North America where land lies below sea level is in Death Valley, California. For data in these locations, the applicable Scale_Factor parameter inserted into the projection file will be a value less than 1, such as 0.999993 for example.

SAVING THE CUSTOM PROJECTION FILE TO DISK

When the best possible alignment of the CAD file has been achieved, go to View > Data Frame Properties > Coordinate System tab, and click Add to Favorites in the lower right corner of the dialog box.

Refer to appendix C for the default location where this custom projection file will be saved. This will depend on the version of ArcGIS Desktop you are using, as well as the operating system of your computer.

If a folder in the path indicated is not visible, go to View > Folder Options > View tab, and change the option button to "Show hidden files and folders". Also uncheck the two boxes immediately below that button. When Windows produces a warning, click Yes on the warning. At the top of the View tab, click Apply to All Folders, then click Apply and OK. The hidden folder will now be visible.

In Windows Explorer, copy the custom PRJ file, then go to the installation location for ArcGIS Desktop on your computer, and open the Coordinate Systems directory. If you have not already created a new folder in Coordinate Systems, do so and name it Custom PRJ Files or another name that is meaningful to you. Paste the custom projection file into the folder.

APPLYING THE CUSTOM COORDINATE SYSTEM TO THE CAD FILE

This step does not work if the CAD file has spaces in the name. If there are no spaces in the name of the CAD file, paste a copy of the PRJ file into the same folder with the CAD file. Rename the PRJ file with exactly the same name as the CAD file. This is case-sensitive. For example, if the CAD file name is Parcels022007-G.dwg, name the PRJ file Parcels022007-G.prj.

If multiple CAD files in the same local coordinate system are stored in the folder, the custom projection file can be renamed "esri_CAD.prj". ArcMap will recognize this coordinate system definition and apply the same coordinate system to all CAD files in the directory.

The projection file named esri_CAD.prj will define the projection for CAD files with spaces in the name.

SUMMARY

The complex issues discussed in chapter 5 merit carefully study if your workflow includes working with CAD data in ArcMap. Unusual units are sometimes used to create the data in CAD programs. This chapter examines techniques for identifying them and for modifying a standard projection file installed with ArcGIS Desktop to incorporate these units. Chapter 5 also discusses the method for applying a grid-to-ground scale factor in the projection in order to align data that represent features above or below sea level. The chapter examines other issues that affect display or alignment of CAD data in ArcMap.

In chapter 6, the process of creating a custom projection file to align CAD data that has been rotated in the native application is explained in detail.

CHAPTER 6

ALIGNING ROTATED CAD DATA

"I received a CAD file from a client, but when I add it to ArcMap it draws at an angle. How can I fix that?"

The two previous chapters, 4 and 5, presented relatively straightforward methods for aligning data in ArcMap. These methods include modifying the false easting and false northing values, identifying and modifying units, and adding a scale factor parameter to the projection file. All these modifications may be needed in addition to compensating for rotation in the CAD file.

ADJUSTING FOR THE AZIMUTH OR ROTATION PARAMETERS

Customizing coordinate systems that include an azimuth or rotation parameter is difficult. These coordinate systems require you to be able to determine precise coordinates for the origin of the rotation. The azimuth or rotation value in the projection file represents a rotation of the data from the point of origin. Imagine sticking a pin in a piece of paper and spinning the sheet. The sheet rotates around the pin. The central meridian and latitude of origin in this operation literally pin the data to the reference data. If the pin (the point where the central meridian and latitude of origin values intersect) is not in the true center of rotation in the local coordinate system, the result will be data that still appears rotated, either clockwise or counterclockwise, when you have finished making adjustments to the projection file. Lining up one point or area within the data will result in data in other areas being offset farther from where the data should be.

If the CAD data has been rotated in the native application, and the rotation appears in the data when added to ArcMap with your reference data, you should ask the CAD operator to remove the rotation from the file for you. The process in this chapter is the most complex task described in the book. The process can be very slow, tedious, and can consume several hours of your time, as well as try your patience.

COMPENSATING FOR THE ROTATION

If removing the rotation in the original application is not possible, here are the steps to compensate for the rotation.

Open ArcMap with a new, empty map, and add the reference data with which the CAD file will align. The reference data must have the coordinate system defined, and must be in a projected coordinate system that uses units of feet or meters, such as state plane. As you will see in the following example, the reference data can even be the polylines from another CAD file, as long as the CAD file has the coordinate system correctly defined. In this case, the coordinate system for the reference data, Index.dwg, is NAD 1983 StatePlane New Hampshire FIPS 2800 Feet.

In this example the rotated CAD file with the unknown coordinate system is named Parcels.dwg. Using the Add Data button again, bring in only the polyline layer from the CAD file. To facilitate data display, the reference data and the CAD file displayed in ArcMap should be stored on the local hard drive rather than over a network.

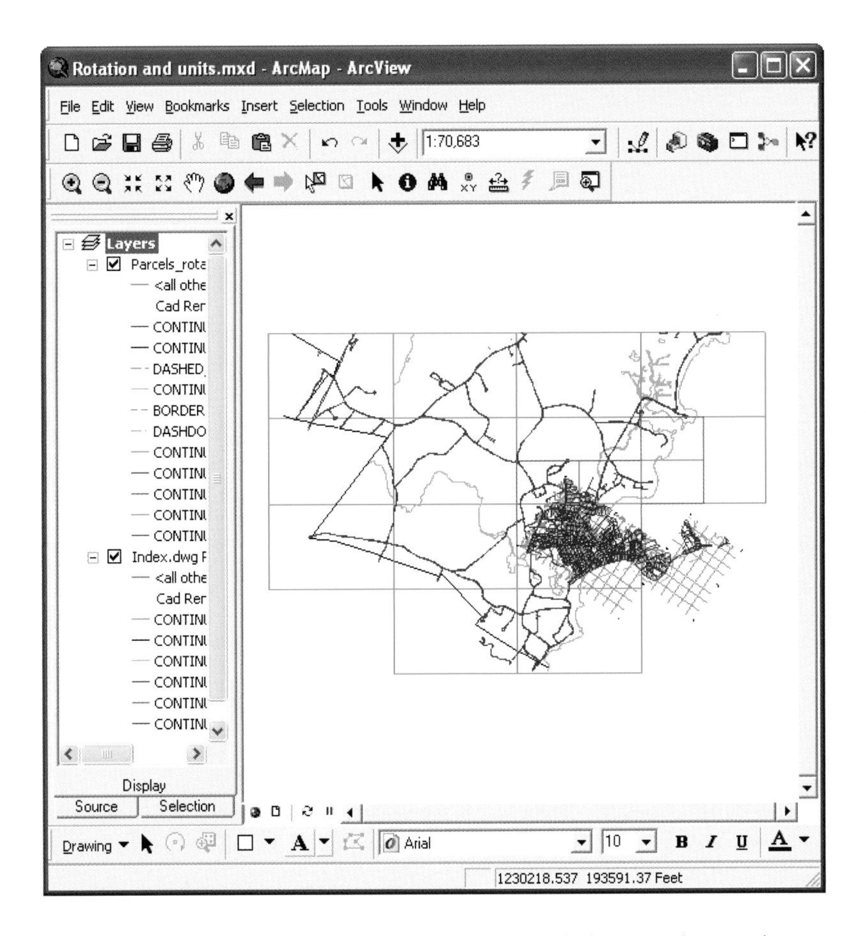

Figure 6–1 Parcels.dwg appears to be in approximately the correct location, but is rotated clockwise in relation to Index.dwg. Parcels.dwg is supposed to fit within the magenta outline that is part of Index.dwg.

While Parcels polylines are in the same location relative to the polylines from the reference layer (Index.dwg), the Parcels data appears rotated in relation to the reference layer.

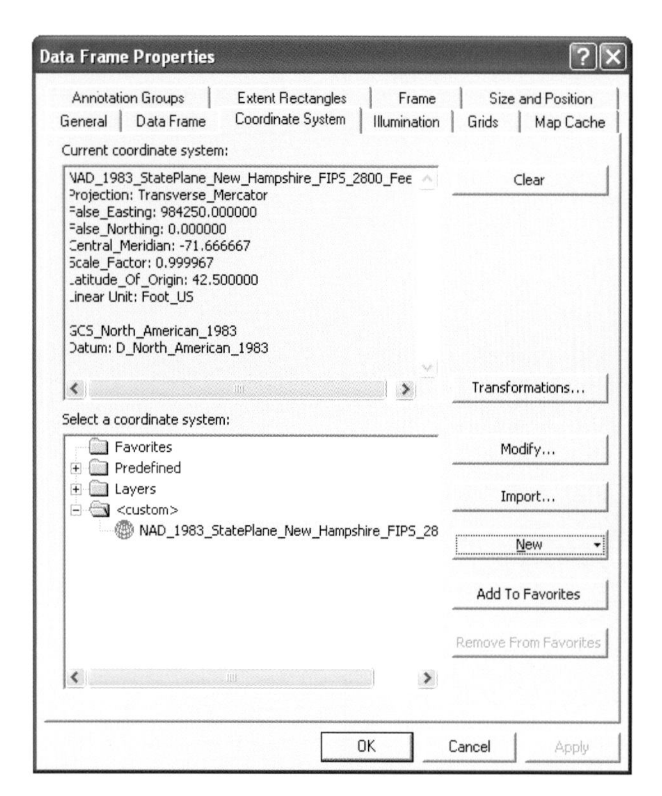

Figure 6-2 Coordinate system of the ArcMap data frame is set to the projection of Index.dwg, the reference data.

In this case, the point of origin in decimal degrees for the CAD file is unknown. To determine reasonable values for the origin of the rotated CAD file, change the Display units of the ArcMap data frame to decimal degrees (as shown in figure 6-3): go to View > Data Frame Properties > General tab. Change the Display units to Decimal Degrees, then click Apply and OK.

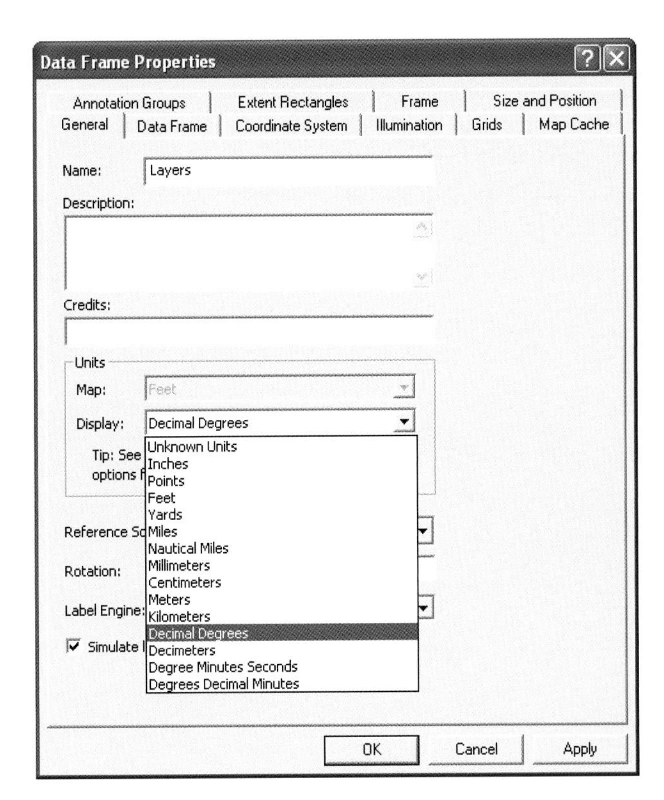

Figure 6-3 Setting Display units to Decimal Degrees allows you to estimate the central meridian and latitude of origin parameters needed to create a custom projection file to line up Parcels.dwg. These parameter values must be entered in the projection file in decimal degrees.

Now the coordinates for the cursor location in the ArcMap status bar will be displayed in decimal degrees. That will enable you to obtain values for the central meridian, standard parallels, and latitude of origin for the custom coordinate system. Moving your cursor across the screen, estimate the origin of the rotation of Parcels.dwg as illustrated in figure 6-4. This point appears to be close to the lower left corner of the magenta outline in Index.dwg.

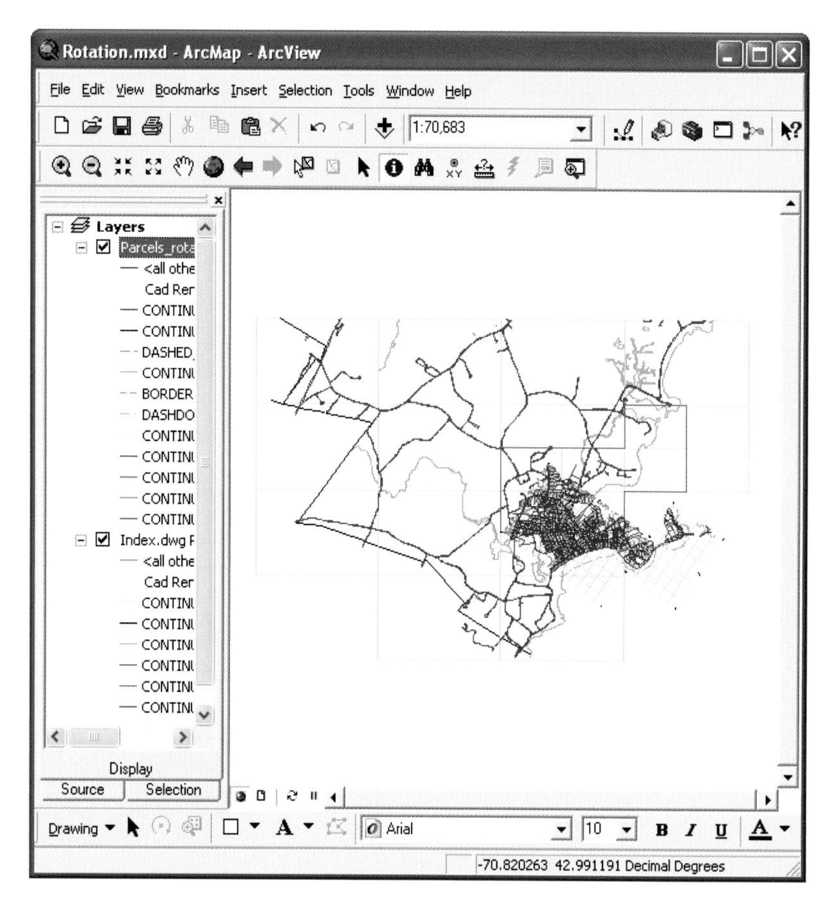

Figure 6-4 The origin of the rotation appears to be the lower left corner of Parcels.dwg, and corresponds to the lower left corner of the magenta box that is part of the reference file, Index.dwg.

The origin for rotation in the CAD program will often be the lower left corner of the drawing. In this case, the coordinates for that location are approximately

X = -70.7744086

Y = 42.9849317

These are the values that will be used for the central meridian (x) and latitude of origin (y) in the custom projection file.

Notice that the Parcels drawing file contains a lot of very detailed data, and it is difficult to see exactly where the data is supposed to align with Index.dwg. Zooming in, you can see the red lines that show the parcels in the data are interfering with the view. Use the Identify tool to determine which layer from the CAD file contains the parcel/red lines.

In this case, the name of the offending layer is LOT-L.

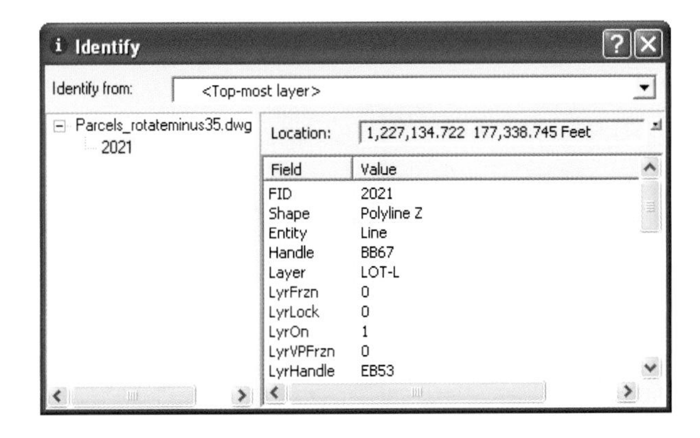

Figure 6-5 The window displaying the results from the Identify tool shows that the red lines in Parcels obscuring your view are in Layer LOT-L.

Right-click the name of the CAD polyline layer (top right in figure 6-5), go to Properties > Drawing Layers, and uncheck the box for the layer that provides too much detail. Click Apply and OK.

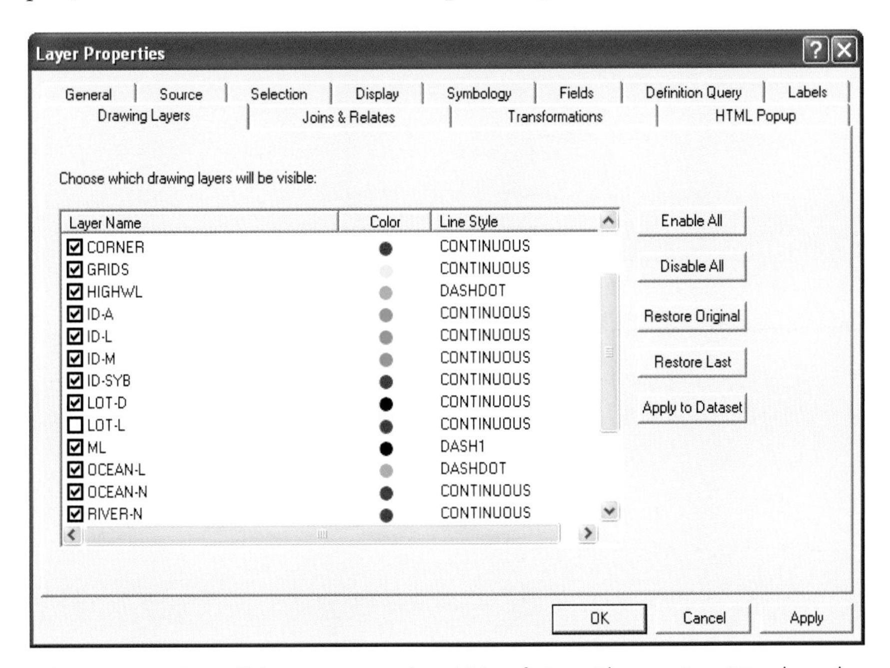

Figure 6-6 Turn off the unnecessary layer(s) interfering with your view. (Not shown here, but also turned off are BLDGS and TREE-L.)

Zooming back to the full extent, you see additional data that is unnecessary for your purposes here: building footprints on Layer BLDGS can be turned off and the Layer TREE-L is not useful either. Removing unnecessary detail allows you to see more clearly whether the data in Parcels.dwg aligns with the data you're using as a reference for comparison, Index.dwg.

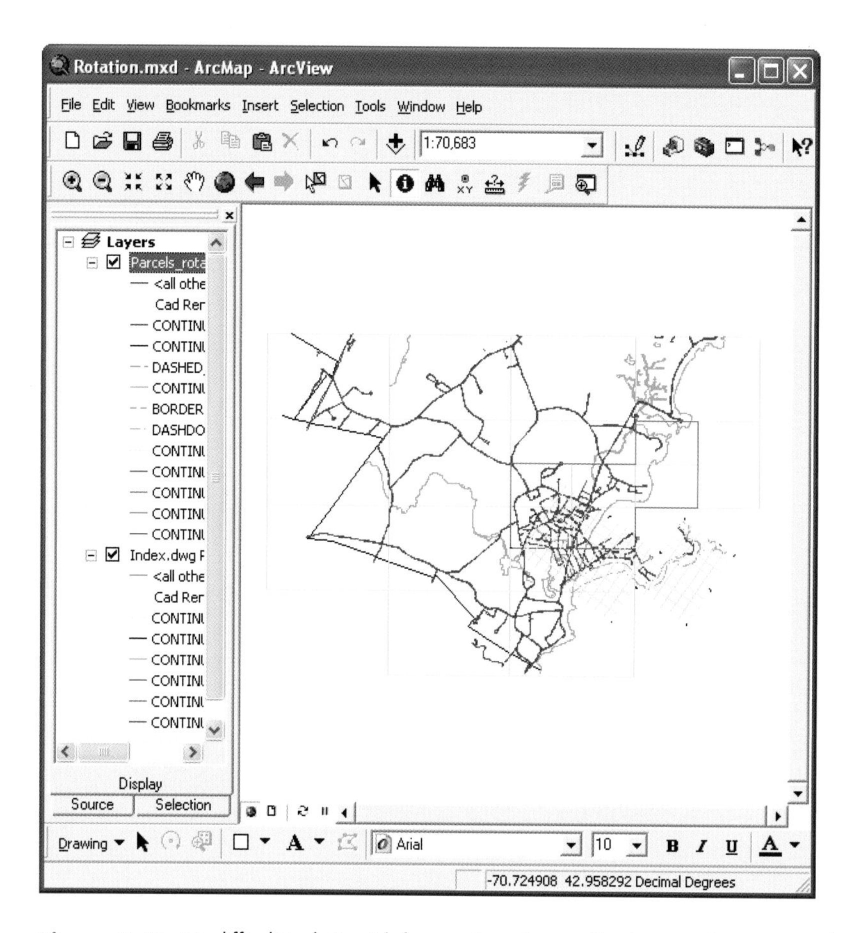

Figure 6–7 It is difficult to distinguish features from the two files because the remaining lines are drawn with the same color.

Now there is another issue. The roads in both Index.dwg and Parcels.dwg are blue, the same color. Change the draw color for one of the files so they contrast. One suggestion is to use blue for the reference data, and red for the data being aligned. Since Index.dwg already has blue roads, change the color of the layers drawn for Parcels to red.

Right-click the name of the Parcels layer, select Properties > Symbology tab. Under Show: select Features > Single Symbol, and change the draw color to red, then click Apply and OK.

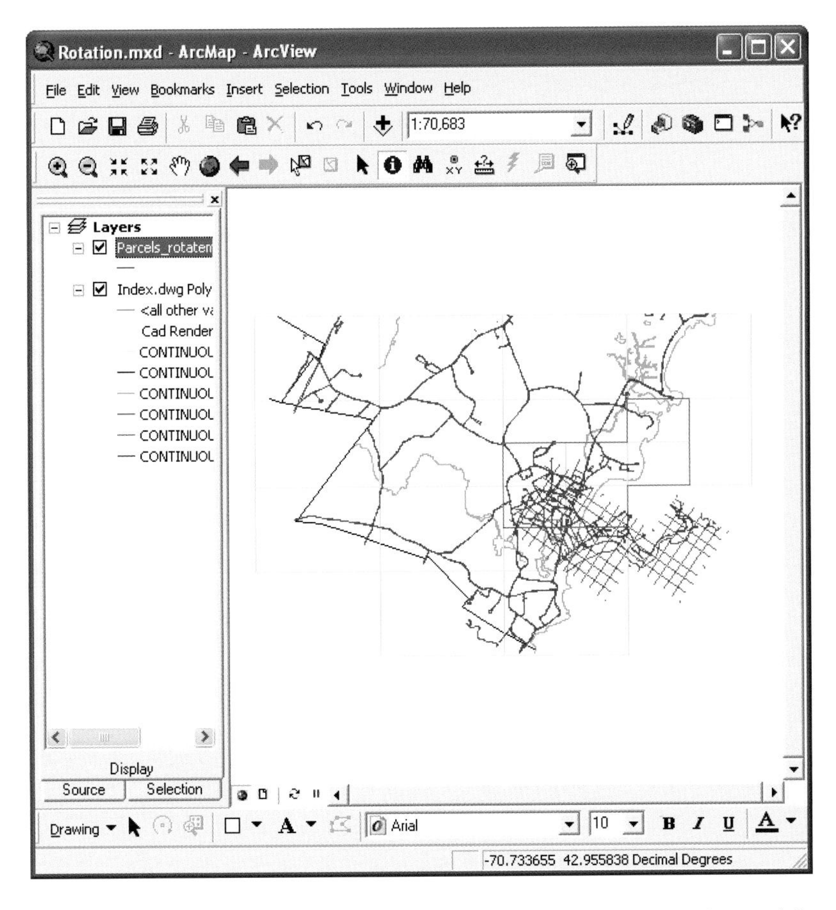

Figure 6–8 Changing the color of one layer (Parcels) makes it easier to distinguish between the two datasets in the map.

Since the data in Parcels appears to be the same size as the magenta box the data is supposed to lie within, you know that the units for the data in Parcels must be feet, matching the units for the data you're using for comparison, Index.dwg.

Now that you have found approximate coordinates in decimal degrees for the center of rotation for Parcels.dwg, and found that the units of measure for its CAD data and the reference data, Index. dwg, are the U.S. survey foot, you are ready to begin creating the custom projection file that will align the data.

CREATING THE CUSTOM COORDINATE SYSTEM

Figure 6–8 not only shows the Parcels data to be the same size as the reference dataset but also makes obvious that the data is rotated. Now that you know Parcels' units are in feet, though, you can accommodate the rotation in the CAD file by creating a custom coordinate system for CAD data in ArcMap. To begin creating the custom coordinate system, go to View > Data Frame Properties > Coordinate System tab. Click New > Projected Coordinate System.

Enter a name for the new custom projection file with no spaces and no special characters.

The following projections supported in ArcGIS Desktop can be used to create custom coordinate systems for CAD data in ArcMap to accommodate rotation in the CAD file:

Local
Rectified skew orthomorphic center
Rectified skew orthomorphic natural origin

It is best practice to select Local from the Projection Name drop-down list and to use this option whenever possible.

Enter the central meridian and latitude of origin collected earlier that appear to be the center of rotation for Parcels. The name of the central meridian parameter in this projection is Longitude_Of_Center. The name of the latitude of origin parameter is Latitude_Of_Center. Remember that in North America, the Longitude_Of_Center value will be negative.

Note that the azimuth parameter has a default value of 45.00. This is the default angle that will be applied to the data. Examining the data in figure 6-1 (page 81), you can see that the rotation angle is not quite as large as 45 degrees. For testing, though, you can leave the azimuth value as it is temporarily, although if you have a protractor available you can measure the angle.

The sign of the azimuth, + or -, is very important. A positive azimuth value will rotate the data clockwise (to the right) in relation to the reference data. A negative azimuth value will rotate the data counterclockwise, to the left, in relation to the reference data. Since Parcels.dwg needs to rotate counterclockwise, the sign for the azimuth value needs to be changed to a negative.

SELECTING A GEOGRAPHIC COORDINATE SYSTEM

Since we do not know if any datum was applied to the CAD file when it was created, we can assign to this new projection file the same geographic coordinate system that was used for Index.dwg.

Click Select > North America folder > North American Datum 1983.prj. The Projected Coordinate System Properties dialog box will resemble figure 6-9.

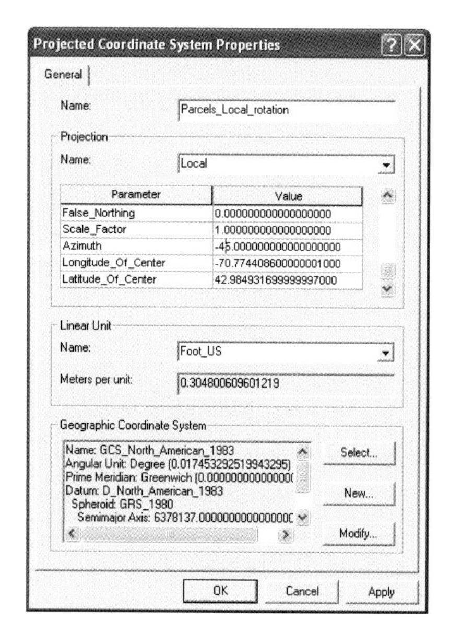

Figure 6-9 The Projected Coordinate System Properties dialog box shows the new values entered for the Local projection.

Click Finish. On the Data Frame Properties dialog box, click Apply, then Add to Favorites, then OK.

Zoom to the full extent. You will see that Parcels data is now a long distance from the Index.dwg polylines. You will need to measure and adjust the false easting and false northing values to line up the Parcel polylines with Index.dwg.

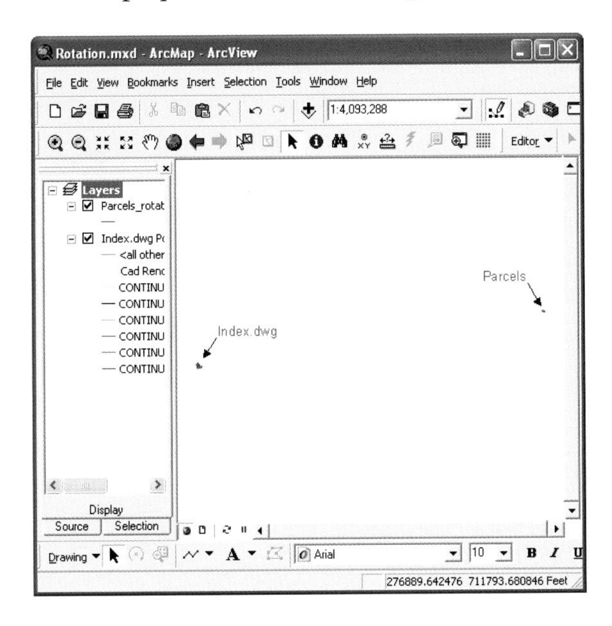

Figure 6-10 Parcels data (red) is now quite a distance from Index.dwg (blue). The false easting and false northing need to be adjusted to move Parcels back in line with the reference data. Parcels.dwg will be moved west by making the false easting value larger.

START MEASURING

Be sure the units in the Measure tool are set to feet, and measure in a straight line directly east to west, since Parcels is offset a greater distance east to west than north to south.

Go to View > Data Frame Properties > Coordinate System tab, and click Modify.

Enter the value from the Measure tool in the false easting value field. The Parcels need to move west, so the new false easting value will entered as a positive number to make it larger.

Click Apply and OK, then Apply, Add to Favorites, and OK on the Data Frame Properties dialog box.

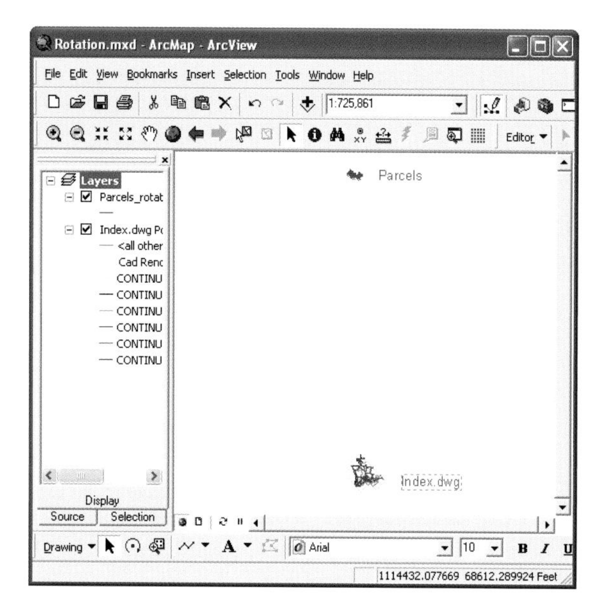

Figure 6-11 Making the false easting value in the projection file larger moved Parcels west to align better with Index.dwg but Parcels is still too far north.

Click Full Extent. The Parcels polylines will be much closer to those of Index.dwg, but Parcels is still north of where the data should be.

Measure the distance in a direct north-south line from Parcels.dwg to Index.dwg.

Go back to View > Data Frame Properties > Coordinate System tab, and click Modify.

The *local* coordinate system originally has a false northing value of 0. Since Parcels needs to move *south*, enter the north-south measurement *as a positive number*.

Click Apply and OK, then Apply, Add to Favorites, and OK on the Data Frame Properties dialog box.

Zoom to the full extent again. The data in Parcels.dwg is now in the same location as the Index.dwg data, but the angle is wrong. The -45° default azimuth value from the original projection file is too large and rotates Parcels.dwg too far in the clockwise direction.

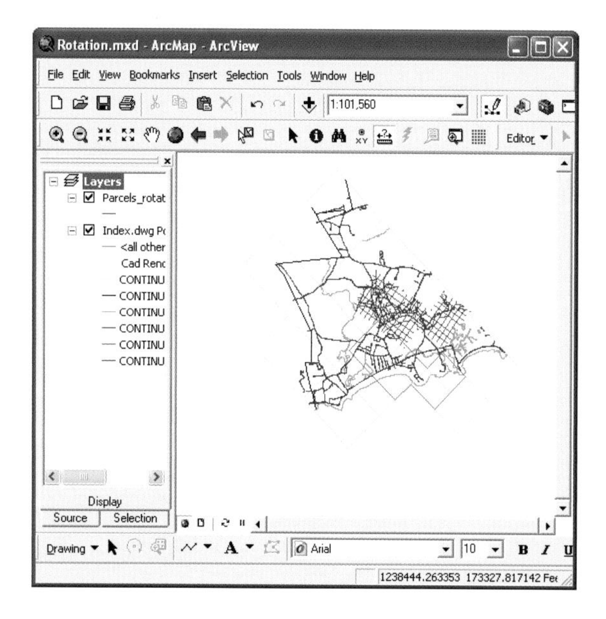

Figure 6-12 Parcels is still too far north, and the rotation angle is not quite right. The azimuth value of -45 degrees needs to be smaller.

Go back to View > Data Frame Properties > Coordinate System tab, and click Modify.

MODIFYING THE AZIMUTH VALUE

In this case, we changed the azimuth from -45 to -35. Changing the azimuth angle will require further adjustments to the false easting and false northing values.

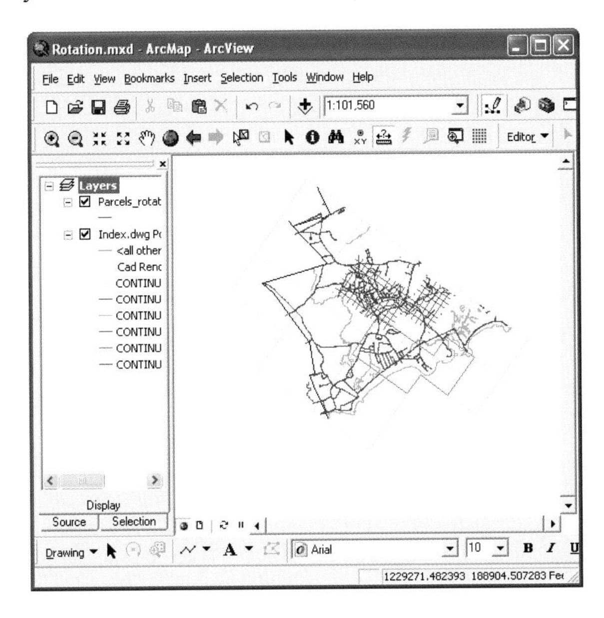

Figure 6-13 Now Parcels.dwg and Index.dwg are at about the same angle, but Parcels needs to move east and south to align exactly.

Now in order for Parcels to move east as it should, the false easting needs to get *smaller*. Parcels should move south, so the false northing needs to get *larger*. (If you need to review why, see chapter 5.)

By your zooming in on the data and continuing to measure offsets and to adjust the false easting and northing, the data alignment improves, but you can see that the azimuth angle is still not quite right.

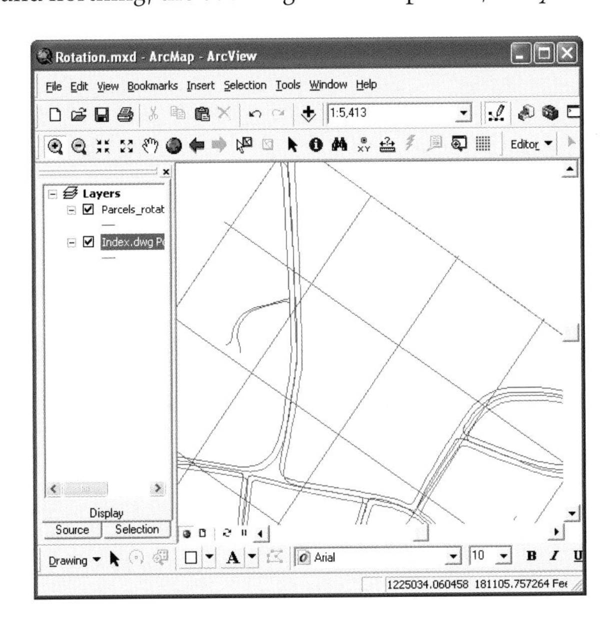

Figure 6-14 The azimuth angle needs additional adjustment since the current angle of -35° is still slightly too large.

Modify the azimuth in small increments, adjusting the false easting and false northing to realign the data each time the rotation is changed.

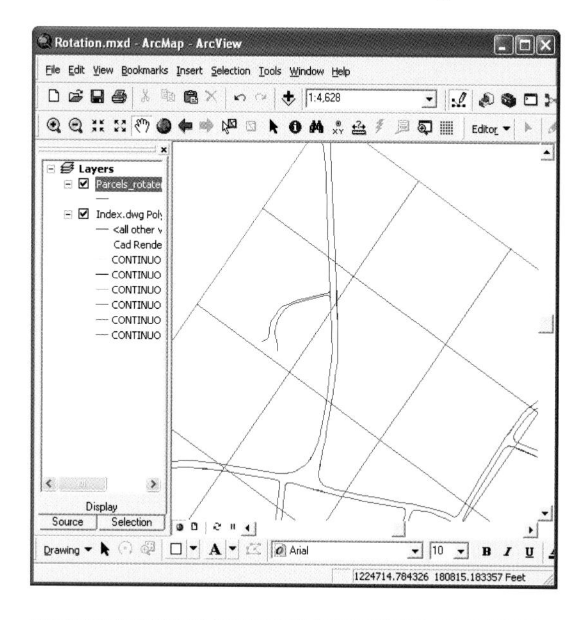

Figure 6-15 At a map scale of 1:4,628 the data alignment looks very good. Zoom in further to check if additional adjustments are needed.

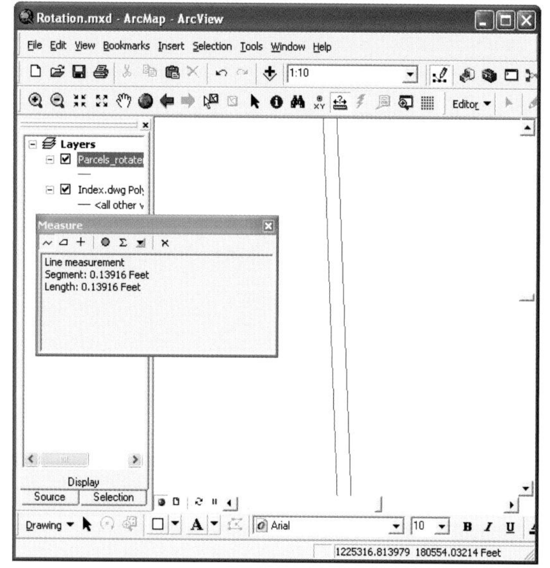

Figure 6-16 One additional tiny adjustment can be made, making the false easting 0.13916 feet larger to move Parcels.dwg west.

At this point, your map scale is 1:10, and the offset between the Parcels lines (red) and the Index.dwg lines (blue) is about 14/100th of a foot. By making one additional adjustment to the false easting, you can see the data line up perfectly in figure 6-17.

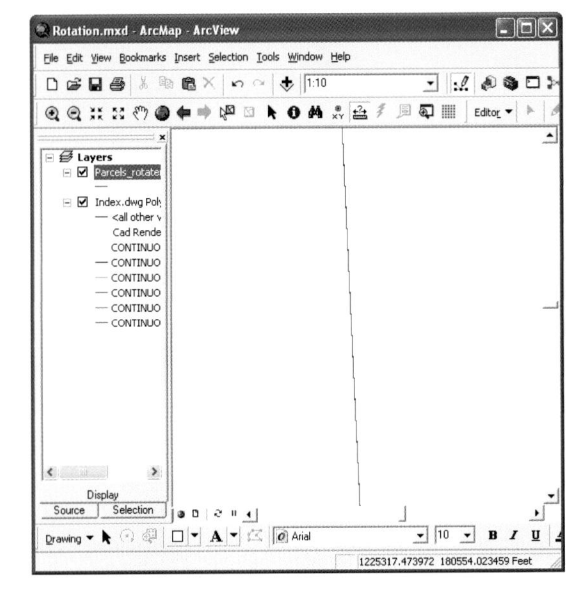

Figure 6-17 With the final false easting adjustment, the data alignment is successful.

As evident in this example, in your own work you will need to test various values for the azimuth in the Local projection in order to determine the correct rotation angle. Since each adjustment of the azimuth will affect the false easting and false northing as well, patience is required in order to perfect the data alignment. Once you have it, the Data Frame Properties dialog box will display the final parameters (as in figure 6-18).

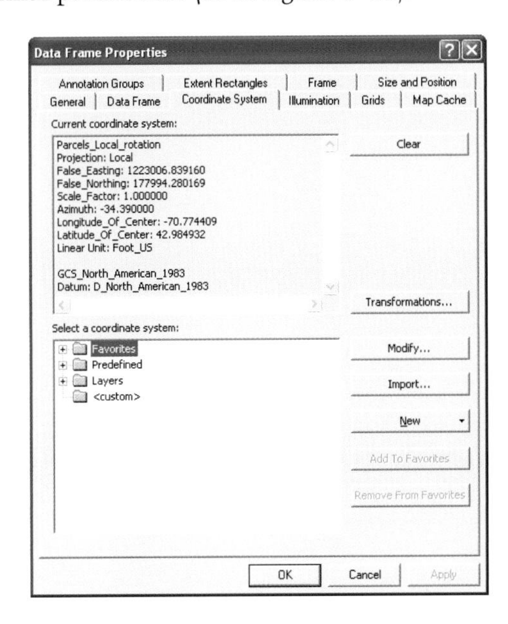

Figure 6-18 Here are the parameters for the final version of the custom projection file for Parcels.dwg created in this example.

Detailed instructions for saving and applying the custom projection file to the CAD data are provided at the end of chapter 4 on page 59–60.

OTHER CUSTOM COORDINATE SYSTEM OPTIONS

As mentioned earlier in this chapter, there are other projections that can be used to create a custom projection file for CAD data that will accommodate a rotation angle: rectified skew orthomorphic center and rectified skew orthomorphic natural origin. The mathematics used to calculate these two projections are different from those used for local, the option that allowed us to achieve nearly perfect alignment of the rotated Parcels drawing file in the previous example.

Rectified skew orthomorphic center and rectified skew orthomorphic natural origin, also known as RSO, include not only an azimuth parameter but also a rotation parameter. Either of these options can be used to correct for rotation in the CAD file. However, adjustment of the azimuth or rotation parameters using the RSO natural origin option moves the data much farther away from the reference data than the same values applied using RSO center. Therefore, using RSO center to align the data is somewhat easier and quicker than RSO natural origin.

Using the same sample files, Parcels.dwg and Index.dwg, figure 6–19 shows the projection parameters that align Parcels.dwg with Index.dwg, using the rectified skew orthomorphic center projection. Compare the projection parameter values shown in figure 6–18 to the local coordinate system. Notice that the values are identical, except that the sign for the azimuth parameter is reversed. The azimuth for RSO is a positive number, while for the local coordinate system, the azimuth parameter value is negative.

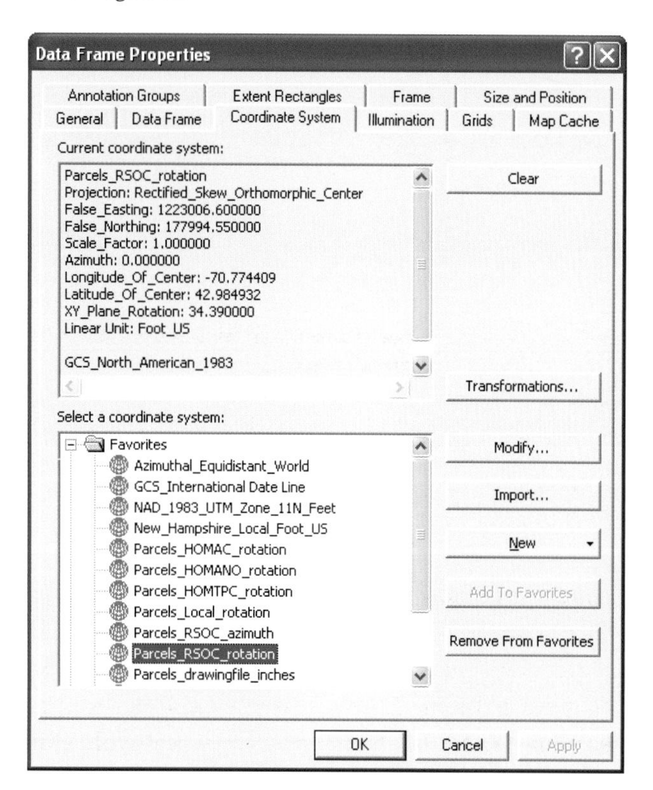

Figure 6–19 Here are the parameters to align the data using RSO_Center. Note that the azimuth value for the local projection shown in figure 6–18 is -34.39, but the rotation parameter here, while it has the same numeric value, is a positive number.

RECTIFIED SKEW ORTHOMORPHIC NATURAL ORIGIN

As mentioned earlier, the rectified skew orthomorphic natural origin projection can also be used. This variation of RSO also includes not only an azimuth parameter, but also the XY_Plane_Rotation. However, the mathematics of this projection dictate that the term *natural origin* in this projection means that the Cartesian origin of the projection is the point where the central line (Longitude_Of_Origin) crosses the aposphere, a special term that refers to a geometric surface. This is almost the ellipsoid's equator. Calculating such a parameter is much better left to the experts.

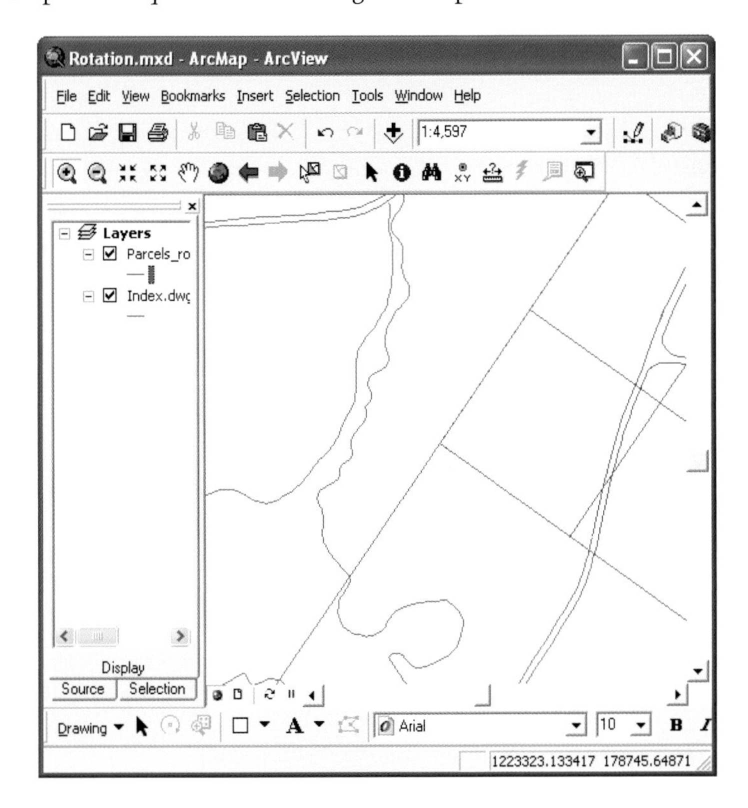

Figure 6-20 Here are the parameters for the final version of the custom projection file for Parcels.dwg created in this example.

When aligning a rotated CAD file, use the local coordinate system option if possible. It will usually provide acceptable results with much less work than the other available projection options.

SUMMARY

Chapter 6 describes how to apply an azimuth and/or rotation parameter to a custom local or RSO projection file and adjust these parameters, together with the false easting and false northing, to align a rotated CAD file with other data in ArcMap. After reviewing this chapter, you will probably agree that having the CAD operator remove the rotation from the file in the native application provides a much easier solution for this issue. Nevertheless, mastering this technique adds a valuable skill to your arsenal as a GIS analyst.

In the next chapter, we take an in-depth look at geographic transformations and the purpose they serve when displaying or projecting data. Chapter 7 also describes the geographic transformation methods that are supported in ArcGIS Desktop, and how custom transformation methods can be applied in the software.

CHAPTER 7

WHY WE NEED GEOGRAPHIC (DATUM) TRANSFORMATIONS

"I have some new data, but when I add it to ArcMap with my other data, a message box pops up that says 'Geographic Transformation Warning' and the data is off by about 200 feet. What am I supposed to do?"

The earth is not perfectly round. Its diameter, measured through the earth at the equator, is longer than the distance from the North Pole to the South Pole. In addition, there are great variations in elevation from the top of Mount Everest (29,029 feet, 8,848 meters) to Challenger Deep in the Marianas Trench (35,827 feet, 10,920 meters) (*National Geographic Atlas of the World, 7th Edition*, 1999).

The spheroid is a model of the earth's surface that averages these irregularities. Based on available measurements for diameters of the earth across the equator and between the poles and for the curvature of the surface, geodesists have calculated many different spheroids to represent an average shape of the earth. A datum is then constructed on a particular spheroid to model the surface of the earth for a specific area, because the selected spheroid "fits" the surface of that area particularly well.

Since so many hundreds of datums are defined for various parts of the world, you will often receive data for the same area on different datums. Because the underlying spheroids on which the datums are based represent the earth with a different size and curvature of the surface, data on different datums will not line up when added to ArcMap.

In order to reconcile these differences, and work with data on different datums, geodesists have calculated geographic transformations between datums. These transformations recalculate coordinates for data from the original datum to a new datum.

The correct geographic transformation must be applied when data on different datums is displayed in ArcMap, or the data will not align properly. Similarly, if the correct geographic transformation is not applied when data is projected to a new geographic coordinate system, using the Project tool in ArcToolbox, the output data will be in the wrong place.

For some areas, more than one geographic transformation may be available between the same datums. The accuracy of these different transformation methods varies. Applying the correct transformation, therefore, may also involve selecting the appropriate transformation from the available options. A complete list of geographic transformations and areas of use is linked to Knowledge Base article 21327 at the ESRI Support Center. Instructions on how you can access the article are in appendix A.

WHAT ARE GEOGRAPHIC TRANSFORMATIONS?

Geographic (datum) transformation methods adjust the coordinates of data, created on one geographic coordinate system (GCS), to another GCS that uses a different spheroid and datum, or a different prime meridian or different angular units. In common usage, people refer to a geographic transformation as a "*datum* transformation" because this element in the GCS is the one most often being changed when performing a geographic transformation. A geographic transformation must be applied when projecting data to a new coordinate system using the Project tool in ArcToolbox, if the input and output coordinate systems are on different datums; if the two coordinate systems have different prime meridians; or the two geographic coordinate systems use different angular units. A geographic transformation must also be applied in ArcMap when data on different GCS are displayed together in the same map.

What is the inverse flattening ratio?

A spheroid is defined either by the semimajor axis, **a**, and the semiminor axis, **b**; or by **a** and the flattening. The flattening is the difference in length between the two axes expressed as a fraction or a decimal. The flattening, ***f***, is calculated as:

$f = (a - b) / a$

The flattening is a small value, so the quantity **1/*f*** is ordinarily used instead. The spheroid parameters for the World Geodetic System of 1984 (WGS 1984 or WGS84) are

a = 6378137.0 meters

b = 6356752.31424 meters

1/*f* = 298.257223563 meters

***f* = (6378137.0 - 6356752.31424) ÷ 6378137 = 0.003352810665559551323529120275403**

1*f* = 1 ÷ 0.00335281066555955 = 298.25722349076033371649600527157

From the ArcGIS Desktop Online Help at

`http://webhelp.esri.com/arcgisdesktop/9.3/index.cfm?TopicName=Spheroids_and_spheres`

If you have data on the North American Datum 1927 (NAD 1927), the GCS has these values:
Datum: North American Datum 1927
Spheroid: Clarke 1866
Semimajor axis: 6378206.4 meters
Inverse flattening ratio: 294.9786982
Units: degree
Prime Meridian: Greenwich (England)

When transforming from NAD 1927 to NAD 1983, you see that GCS North American 1983 (NAD 1983) has the following values:
Datum: North American Datum 1983
Spheroid: GRS 1980
Semimajor axis: 6378137 meters
Inverse flattening ratio: 298.257222101
Units: degree
Prime Meridian: Greenwich (England)

Because these two spheroids, Clarke 1866 and GRS 1980, have different semimajor axes and inverse flattening ratios, the surfaces of the two spheroids have different curvatures. The datums defined on those curved surfaces also curve somewhat differently. The geographic (datum) transformation adjusts for that difference, so that when data is transformed the output location of the data is correct on the new datum and spheroid.

When performing the transformation from NAD 1927 to NAD 1983, you are changing the datum and spheroid on which the datum is based. The angular units (degrees) and prime meridian (Greenwich) remain the same.

Note that in ArcGIS Desktop, geographic transformations are programmed to work in either direction: in other words, you use the same transformation name for both "to" and "from." The software was designed this way because there are hundreds of transformations to be supported, and it is more efficient to provide one transformation name (such as NAD_1927_To_NAD_1983_NADCON, for use between NAD 1927 and NAD 1983 within the forty-eight contiguous states) and design the software so that the transformation works in either direction.

Every geographic transformation applies to a specific area. A transformation may apply to an area as small as a single island, or a single political division within a county; or to a single country, a continent, or the entire world. To determine the correct transformation to use for your data, you must know the area of use for the transformation.

It is also very important to note that in some areas of the world, several geographic transformations that use different transformation methods or parameter values may be available. These different transformations may have very different levels of accuracy. To select the correct geographic transformation for your area of interest, it is critical that you research the available transformations and select the one that is most appropriate.

GEOGRAPHIC TRANSFORMATION PARAMETER SOURCES IN ARCGIS DESKTOP

The majority of information in the ArcGIS projection engine, including most geographic transformation parameters, comes from the EPSG Geodetic Parameter Dataset. EPSG is the acronym for European Petroleum Survey Group. This organization became part of Oil and Gas Producers (OGP) and is officially called the Geodetics Subcommittee of the OGP Surveying and Positioning Committee. For many years, the OGP has compiled coordinate system-related information for its members, and makes the information available in a Microsoft Access database. The database is particularly useful because it includes accuracy estimates for specific geographic transformations if these were provided by the source agency.

The current version of the database, plus older, superseded versions, can be downloaded free at **http://www.epsg.org**. There is also an online repository at **http://www.epsg-registry.org**.

ArcGIS Desktop version 9.3/9.3.1 supports most transformation methods and other geodesy-related information from version 6.14 of the EPSG database, and also includes information obtained from older versions.

Other geographic transformation parameters in ArcGIS Desktop are provided to ESRI by national government agencies tasked with calculating geodetic data and supplying services for their respective countries. One such agency is the National Geodetic Survey (NGS) of the United States.

GEOGRAPHIC TRANSFORMATION METHODS

Many geographic transformations for data and areas outside the United States have been calculated, and some of these transformations are not supported in ArcGIS. When working with international data, you may be given a geographic transformation method and parameters. In those cases, it is essential to know which geographic transformation methods are supported, be familiar with their names, and how to set up and save custom transformations using ArcGIS.

The following geographic transformation methods are supported in ArcGIS Desktop:

Geocentric translation
Molodensky
Abridged Molodensky
Position vector
Coordinate frame
NADCON
HARN
NTv2
Molodensky-Badekas
Longitude rotation

ArcGIS Desktop Help and many technical papers are available online that describe the mathematics of these transformation methods. Such treatments are beyond the scope of this book, but a brief description of these transformation methods is in order.

THREE-PARAMETER TRANSFORMATION METHODS

Geocentric translation, Molodensky, and abridged Molodensky are all colloquially known as three-parameter transformation methods. The geographic transformation parameters for each of these consists of:

x-axis translation
y-axis translation
z-axis translation

These transformations shift the data by the values entered for the transformation, in the x, y, and z directions, using units of meters. The mathematics of these transformation methods are very similar, but the output results are slightly different. The values are applied after converting the input data's longitude and latitude values into a 3D geocentric system (x,y,z). If necessary, data that is defined with a projected coordinate system will be internally converted in the software to the corresponding GCS, transformed, then converted back to the PCS.

Parameters for these transformation methods can be either positive or negative numbers. For example, the geocentric translation named AGD_1966_To_WGS_1984 is applied to the entire country of Australia. This transformation has the following parameters:

x-axis translation -133.0
y-axis translation -48.0
z-axis translation +148.0

This transformation is supported in ArcGIS Desktop, and the software will apply the transformation in the correct direction, whether you are transforming from AGD 1966 to WGS 1984 or from WGS 1984 to AGD 1966.

However, if you obtain parameters for a geocentric translation from another source, the source must also tell you the *direction in which the transformation parameters are to be applied*. If the transformation was applied to transform from WGS 1984 to AGD 1966, the *signs of the parameters would be reversed* to look like this:

x-axis translation +133.0
y-axis translation +48.0
z-axis translation -148.0

This is a critical point, worth repeating: If you obtain parameters for a geographic transformation that is not supported in ArcGIS Desktop, the source of the transformation must also tell you the direction of the transformation, or the output from the transformation process will be wrong.

SEVEN-PARAMETER TRANSFORMATION METHODS

The position vector and coordinate frame methods are both seven-parameter transformations, with the following parameters:

x-axis translation
y-axis translation
z-axis translation
x-axis rotation
y-axis rotation
z-axis rotation
scale difference

If a source provides seven parameters for a geographic transformation, the source *must* also provide the transformation method and the direction of the transformation.

WHAT DIFFERENCE DOES THE TRANSFORMATION METHOD MAKE?

The transformation AGD_1966_to_WGS_1984_15, supported in ArcGIS Desktop, is used for the Northern Territory in Australia. This is a coordinate frame transformation, with the following parameter values:

x-axis translation +124.133
y-axis translation -42.003
z-axis translation +137.400
x-axis rotation +0.008
y-axis rotation -0.557
z-axis rotation -0.178
scale difference -1.854

These exact parameter values are also used for the position vector transformation method, but the signs + and - for the rotation parameters are reversed. Using these parameters to perform a position vector transformation, from AGD 1966 to WGS 1984, instead of applying the coordinate frame method, the parameters would look like this:
x-axis translation +124.133
y-axis translation -42.003
z-axis translation +137.400
x-axis rotation -0.008
y-axis rotation +0.557
z-axis rotation +0.178
scale difference -1.854

Applying the wrong transformation method—or using incorrect signs for the rotation parameters—will result in output data that is in the wrong place.

WHAT DIFFERENCE DOES THE DIRECTION OF THE TRANSFORMATION MAKE?

The parameters listed below, with the positive and negative parameter values as shown, transform from AGD_1966_To_WGS_1984, using the coordinate frame method.
x-axis translation +124.133
y-axis translation -42.003
z-axis translation +137.400
x-axis rotation +0.008
y-axis rotation -0.557
z-axis rotation -0.178
scale difference -1.854

To reverse the transformation and transform from WGS 1984 to AGD 1966—in the other direction, all the signs (+ and -) for the parameters can be reversed. However, this is an approximation. You should define a custom transformation in the same direction as it is given and let the software apply the transformation in the opposite direction if needed.

For supported geographic transformations in ArcGIS Desktop, the source code for the projection engine handles this sign reversal, and you do not have to be concerned about it. But when entering a new custom geographic transformation, you have to know the direction in which the transformation is working, or the signs for the parameters will be wrong and the output data will be transformed in the wrong direction. When the custom transformation has been entered using the Create Custom Geographic Transformation tool in ArcToolbox, the software will also apply the custom transformation in either direction.

NADCON AND HARN TRANSFORMATION METHODS

These transformations are grid-based methods performed using transformation files installed with ArcGIS Desktop, and are applied within the United States and U.S. territories. (Note that the term "grid" used in this chapter refers to a geographic transformation grid file; this "grid" has no relation to the grid raster format commonly used in GIS.) A transformation grid for each of these methods has been created by the applicable agency. The transformation grid will cover a specific area of the earth's surface, and will transform data within that area from one GCS to another. You must select the correct transformation grid for the area in which the data is located. For data outside the extent of the transformation grid, no geographic transformation will be applied to the data.

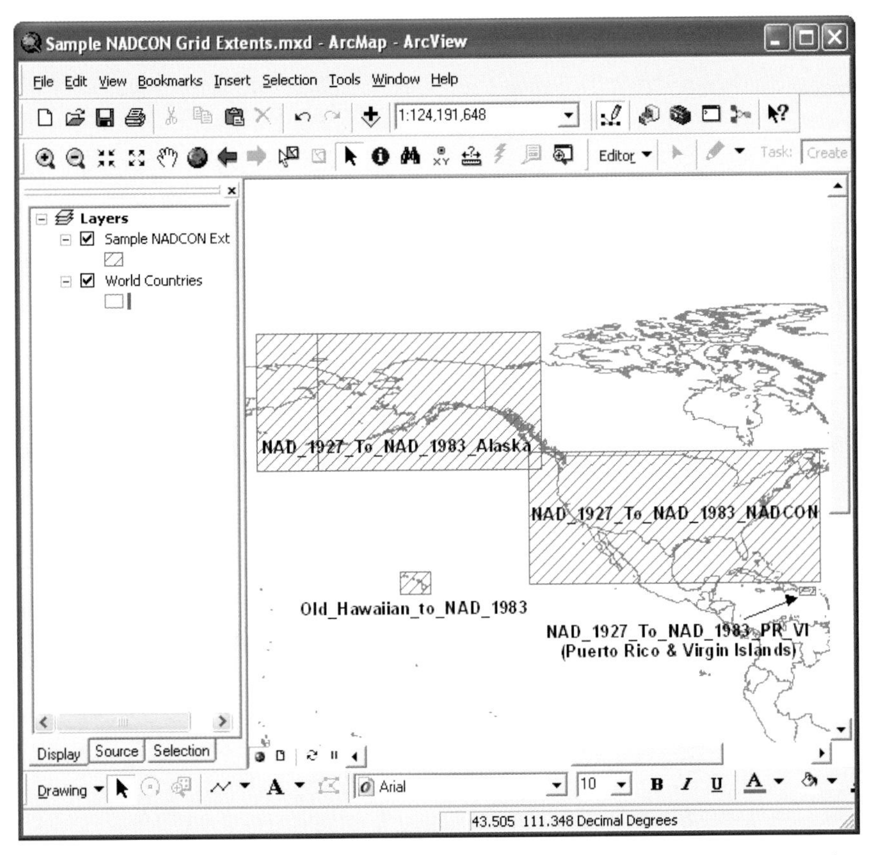

Figure 7-1 Extents for some of the NADCON transformation grids that are supported in ArcGIS Desktop. The extents of additional available grids are too small to display at this map scale.

Figure 7-1 illustrates the extents of some NADCON transformation grids. The accuracy of the NADCON transformation method is 15 centimeters (0.15 meters) at the 67 percent confidence level within the contiguous United States covered by the NADCON grid specified. Accuracies differ in Alaska, Hawaii, and the U.S. territories; check with the National Geodetic Survey for this information. For further information about NADCON and HARN transformations refer to **http://www.ngs.noaa.gov/TOOLS/Nadcon/Nadcon.html**.

High Accuracy Reference Network (HARN) grids transform data over a smaller area, between the NAD 1983 datum and HARN for the specified area. In most cases, a specific HARN grid transforms data for only a single state. In some cases, two or more states will share a single HARN grid; examples include the New England states (Connecticut, Massachusetts, New Hampshire, Rhode Island, and Vermont), Maryland and Delaware, Montana and Idaho, and Washington and Oregon.

Because of their size, Texas and California each have two HARN grids that slightly overlap in the center of their respective states. The HARN grid that applies to both Montana and Idaho is also split into two parts.

The accuracy of the NAD 1983 to HARN transformations is 5 centimeters (0.05 meters) at the 95 percent confidence level.

HARN transformation grids that have been published by the U.S. National Geodetic Survey are installed with ArcGIS Desktop. Alaska is the only state that does not have a HARN transformation grid at this time.

Remember this essential point about grid-based datum transformations: to ensure correct display of the data in the ArcMap data frame and correct output of projected data, when transforming data you must select and apply the correct transformation grid for the area in which the data is located.

NTV2 TRANSFORMATIONS

NTv2 transformations are also grid-based, and are currently used in Canada, Australia, New Zealand, Spain, France, and Japan. While NADCON and HARN grids have the same density for the entire area covered by the transformation grid, NTv2 grids can include higher density "subgrids" that provide a higher level of accuracy for some areas in the grid extent, usually in areas with greater population density, where greater transformation accuracy is required.

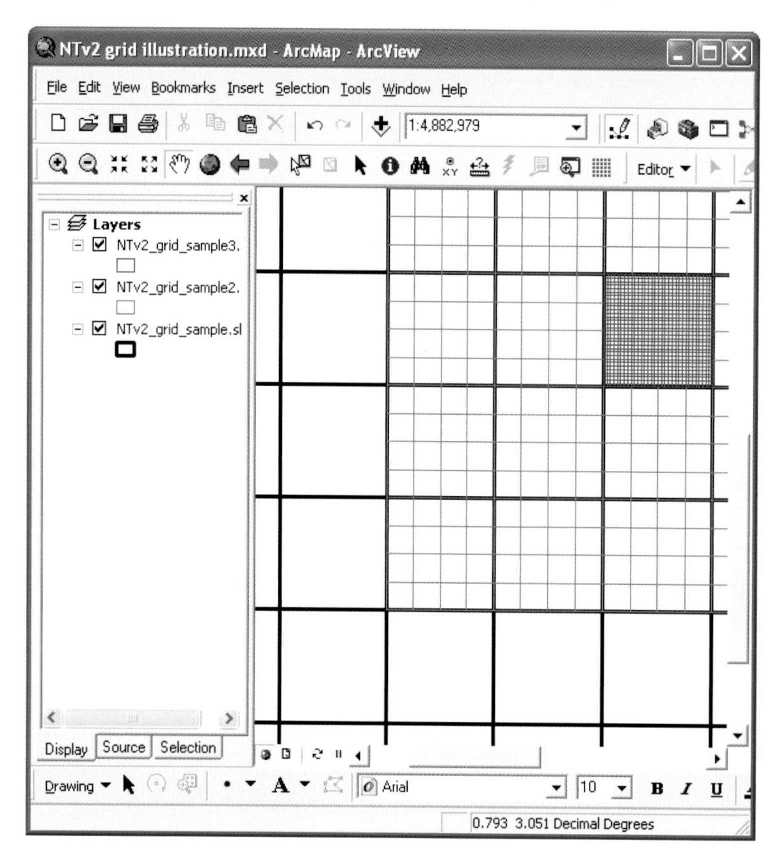

Figure 7-2 The construction of an NTv2 transformation grid. The area in red shows the densest grid where the most accurate transformation will be applied. The areas in blue, with slightly lower population density, have a more widely spaced grid and a slightly less accurate transformation applied. The areas bordered in black show the base grid with the largest "cell" size, where the most general transformation is applied to data falling in those areas.

When the NTv2 transformation is applied, the position of features is checked relative to the grid extent to determine which transformation within the grid will be applied to the data. The features within the area of the greatest density will have the most accurate transformation applied. In general, the maximum published accuracy of the NTv2 grid, within the highest density grid, is 0.01 meters. (The information regarding transformation accuracy and source is published in the EPSG databases.)

You will probably have to install NTv2 transformation grids

Most supported NTv2 transformation grids are not installed with ArcGIS Desktop. In most cases, the required NTv2 transformation grid file, which will have a .GSB extension, must be obtained from the applicable government agency and installed in the correct location in the ArcGIS install folder. If not, in those cases data will not be transformed when these transformation methods are applied. Installation instructions for NTv2 grids are available in Knowledge Base article 35152 from the ESRI Support Center. A list of NTv2 transformations that are supported in ArcGIS Desktop is provided in Knowledge Base article 18317. Instructions on how to access both articles are in appendix A.

MOLODENSKY-BADEKAS TRANSFORMATION METHOD

This transformation is a ten-parameter method that includes the seven parameters used by position vector/coordinate frame methods, as well as three additional parameters. This transformation method is used to transform data between a local datum (with the anchor point or datum origin defined on the surface of the earth) to an earth-centered datum (such as WGS 1984).

x-axis translation
y-axis translation
z-axis translation
x-axis rotation
y-axis rotation
z-axis rotation
scale difference
x-coordinate of rotation origin
y-coordinate of rotation origin
z-coordinate of rotation origin

LONGITUDE ROTATION METHOD

This simple transformation method, true to its name, applies a rotation to the longitude coordinates of the data, to transform the data between two prime meridians, e.g., between Greenwich, England, and Paris, France.

CREATING AND SAVING A CUSTOM GEOGRAPHIC TRANSFORMATION IN ARCTOOLBOX

From a data source, you may obtain the method and parameters for a geographic transformation that is not currently supported in ArcGIS Desktop. Then you can create and save the custom geographic transformation for use in ArcMap and in the Project tool in ArcToolbox: use the Create Custom Geographic Transformation tool located in ArcToolbox > Data Management Tools > Projections and Transformations. (To access this path using ArcGIS Desktop 10, from within ArcMap select Catalog window > Toolboxes > System Toolboxes.)

TO CREATE THE NEW GEOGRAPHIC TRANSFORMATION, YOU MUST BE GIVEN THE FOLLOWING INFORMATION:

1. **Direction of the transformation parameters.**

   In this example illustrated by figure 7-3, a custom transformation for Portugal will be entered. Do the transformation parameters transform *from* Lisbon Hayford *to* ETRS 1989 IGP 2008 or *from* ETRS 1989 IGP 2008 *to* Lisbon Hayford?

2. **The transformation method, from those listed and described earlier in this chapter.**

   As mentioned earlier, the transformation method, together with the direction and +/- signs for the transformation parameters, must be entered correctly or the data will be transformed in the wrong direction and not align correctly.

3. **The transformation parameters with their correct signs.**

   When you have obtained this information, open Create Custom Geographic Transformation in ArcToolbox.

   Enter a name for the custom geographic transformation in that dialog box. The name must not include spaces. Enter the name in the standard format, as in the example below. The name for the new custom geographic transformation must be *unique*—be sure this name is not identical to the name of an existing transformation in the software.

   Name the custom transformation with the proper direction. In this case, parameters were received for a position vector transformation from Lisbon_Hayford_To_ETRS_1989_IGP_2008. Figure 7-3 displays the transformation name and other options to complete.

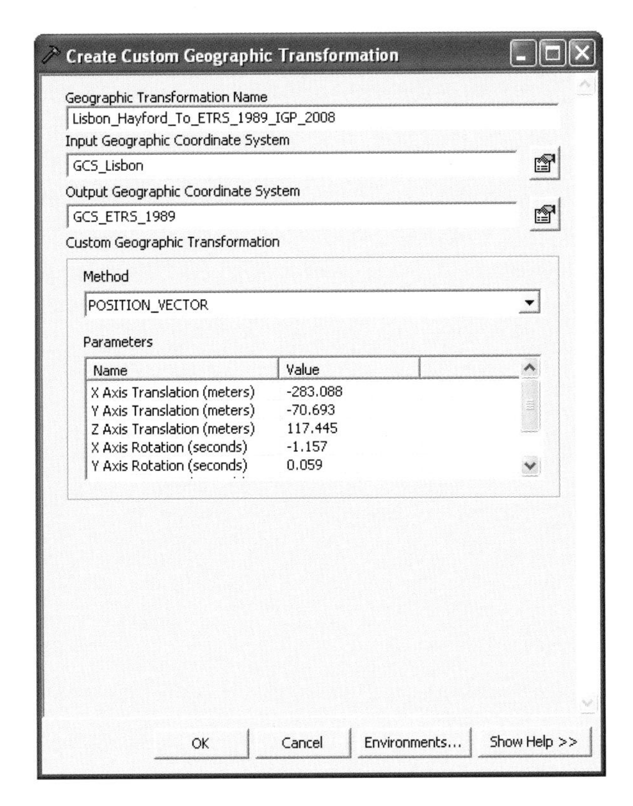

Figure 7-3 The Create Custom Geographic Transformations dialog box. The final parameter, scale factor, is not displayed in this screenshot, but the value for this parameter must also be entered in the dialog after scrolling down to display that parameter and its value box.

The Input Geographic Coordinate System will be Lisbon, so browse to that GCS and select it. (Note that the Hayford spheroid, on which this Lisbon datum is based, is identical to international 1909 as well as international 1924; all three spheroids are based on the Bessel 1841 spheroid and have the same axes measurements and the same curvature.)

The Output Geographic Coordinate System will be ETRS 1989, so select that output GCS.

Select POSITION_VECTOR from the Method drop-down list.

Enter the parameters, *with the correct signs*, provided by the source.

When the parameters have been entered, click OK.

The file named "Lisbon_Hayford_To_ETRS_1989_IGP_2008.gtf" will be saved at a path that depends on your installed version of ArcGIS Desktop and your computer operating system. Refer to appendix C for the specific output location of the GTF file.

After creating the custom transformation, both ArcCatalog and ArcMap must be closed on the computer. When the application is opened again, the custom geographic transformation will be available in the Geographic Transformations drop-down list box in the Project tool, or in the Geographic Coordinate System Transformations dialog box in ArcMap.

After creating the GTF file on one computer, the file may be transferred to the same location on other computers in your office, and will also be available to transform data on those machines.

SUMMARY

Chapter 7 offers detailed information about geographic transformations, vital to know, especially if you work with international data. You should be familiar with the names of the geographic transformation methods supported in ArcGIS, including their parameters and how to set up and save custom transformations. Transforming data is an important part of ensuring correct alignment of the data in the ArcMap data frame and correct output of projected data and this chapter shows you how.

The next chapter also includes instructions that will help you avoid problems in lining up data: chapter 8 focuses on applying geographic transformations in the ArcMap data frame and in the Project tool in ArcToolbox.

CHAPTER 8

APPLYING GEOGRAPHIC TRANSFORMATIONS

"When I add data to ArcMap, and try to set the datum transformation, I get eight different options. Which one am I supposed to pick?"

When adding data to the ArcMap data frame, or projecting data with the Project tool in ArcToolbox, warnings are frequently displayed about different geographic coordinate systems. Because the warnings do not explain what needs to be done to fix the problem, the temptation is to ignore them, but you must not. They are very important, particularly when a geographic transformation is required. This chapter examines these warnings and explains what needs to be done in each case, in order to align data correctly:

- when the data on different GCS is displayed together in ArcMap
- when projecting data between different geographic coordinate systems and datums.

APPLYING GEOGRAPHIC TRANSFORMATIONS IN THE ARCMAP DATA FRAME

In this example, we are using Neighborhoods_sp83.shp. Note that the shapefile name includes some information about the coordinate system of the data—"_sp83" tells us that the shapefile is projected to the state plane coordinate system on the NAD 1983 datum. Adding this shapefile to a new ArcMap session sets the coordinate system of the ArcMap data frame to the coordinate system of the data, which is NAD 1983 StatePlane California VI FIPS 0406 Feet, and it sets the GCS to GCS_North_American_1983.

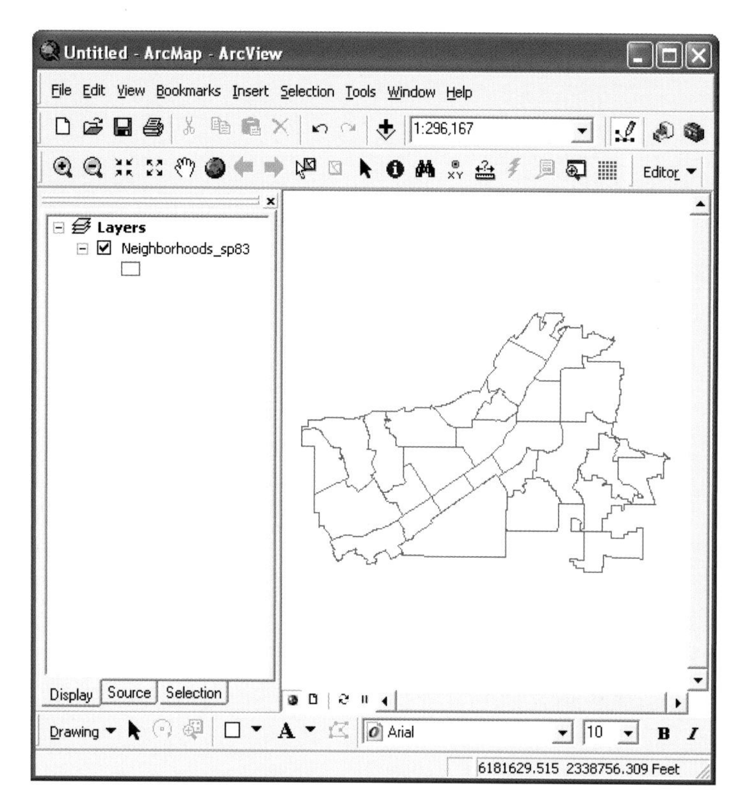

Figure 8-1 Neighborhoods_sp83 projected to NAD 1983 StatePlane California VI FIPS 0406 Feet. Note that the units displayed in the status bar are Feet.

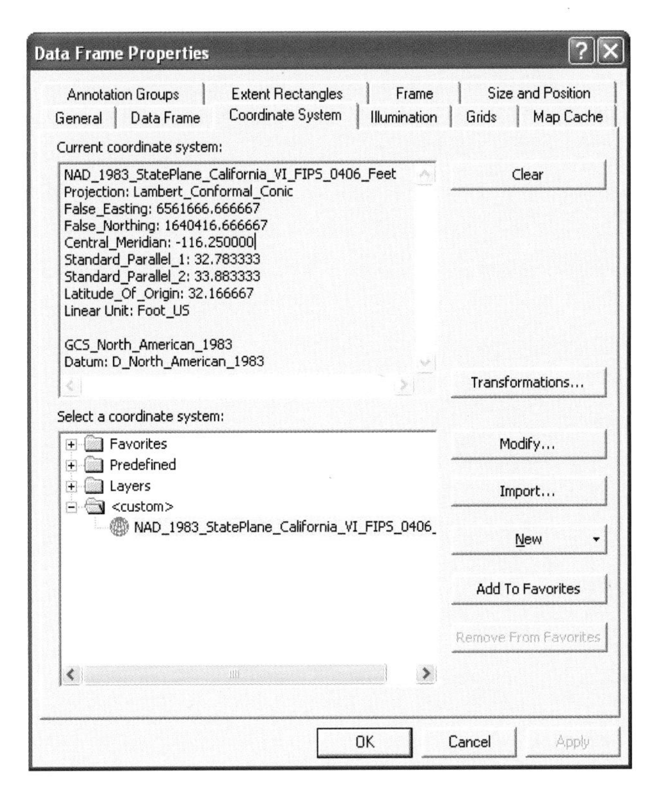

Figure 8-2 The Coordinate System tab in the Data Frame Properties dialog box displays the coordinate system of the ArcMap data frame.

Adding a new shapefile that has been projected to NAD 1983 HARN StatePlane California VI FIPS 0406 Feet to ArcMap triggers the Geographic Coordinate Systems Warning dialog box, shown in figure 8-3.

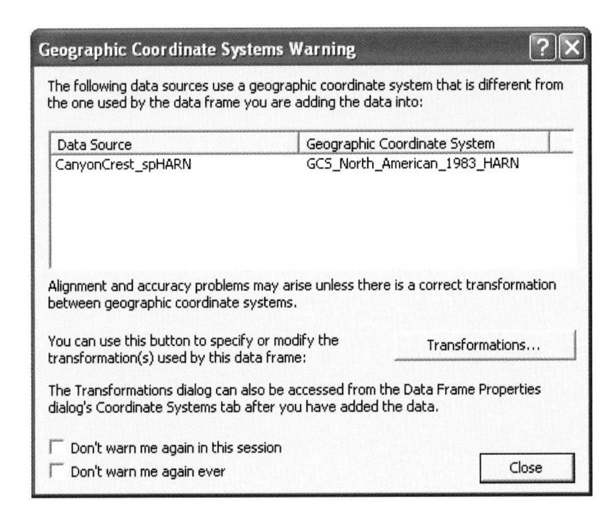

Figure 8-3 This warning notifies you that the data being added to the ArcMap session is on GCS_North_American_1983_HARN. This does not match the geographic coordinate system of the ArcMap data frame, which is set to GCS_North_American_1983.

HARN is a refinement of the North_American_Datum_1983 that is used for highly accurate survey grade data. The offset between NAD 1983 and HARN is very small, but the geographic transformation must be set to align the data correctly. In the Geographic Coordinate Systems Warning dialog box (figure 8-3), you must click Transformations to open the Geographic Coordinate System Transformations dialog box shown in figure 8-4.

No geographic transformations are applied in ArcMap by default. The correct geographic transformation must be selected in the dialog (shown in figure 8-4) or no transformation will be applied to the ArcMap data frame.

There is one exception to the very important statement above. The NAD_1927_To_NAD_1983_NADCON transformation is loaded automatically in ArcMap when data on the NAD 1927 and NAD 1983 datums are added to the same map document. This transformation is used in the contiguous United States—the lower forty-eight states. If data in the map is in a different area, such as Canada or Alaska, the transformation method must be changed so that the correct transformation method is applied.

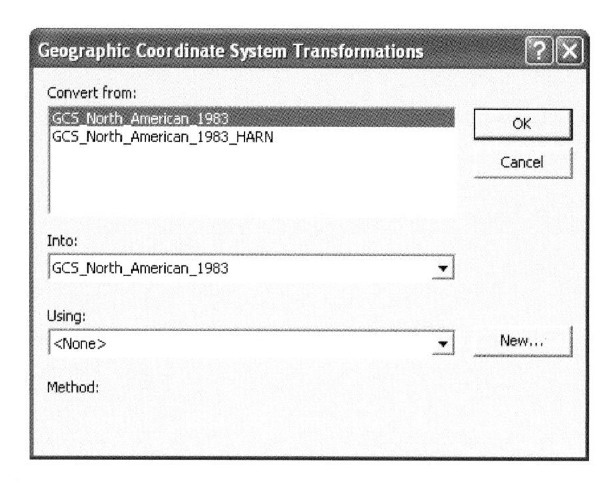

Figure 8-4 Geographic Coordinate System Transformations dialog box. Notice that in the "Convert from" box, the GCS for the two datasets in the map document are listed in alphabetical order. The other GCS listed in the "Convert from" box must be selected manually in order to select the correct transformation from the Using drop-down list.

Figure 8-4 shows that in the "Convert from" box at the top of the dialog, the GCS for the data in the ArcMap document are arranged in alphabetical order. The GCS at the top of the list is selected by default, but that GCS is the same as the one shown in the Into box, so the Using box remains blank.

Because the GCS selected in the "Convert from" box is the same as that in the Into box, the coordinate system of the ArcMap data frame, no transformation methods are available when clicking on the Using drop-down list shown in figure 8-5.

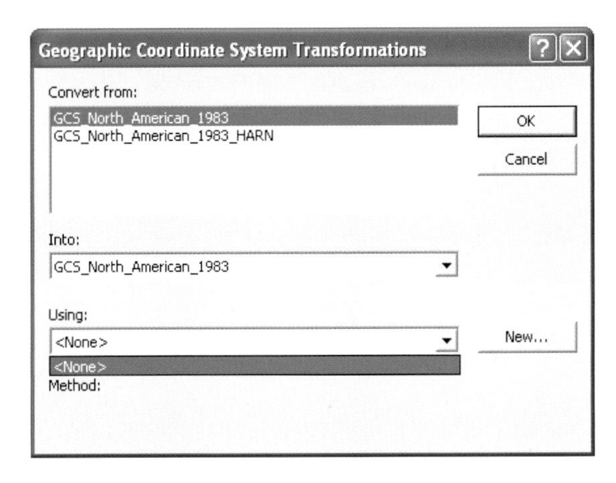

Figure 8-5 No geographic transformations are available unless the GCS to be converted from is selected in the "Convert from" box in the dialog.

Because we are converting from GCS_North_American_1983_HARN to GCS_North_American_1983, the HARN option must be selected in the "Convert from" box to open the list of available transformations. Figure 8-6 shows a partial list of the NAD_1983_To_HARN transformations supported in ArcGIS Desktop for various states or portions of states.

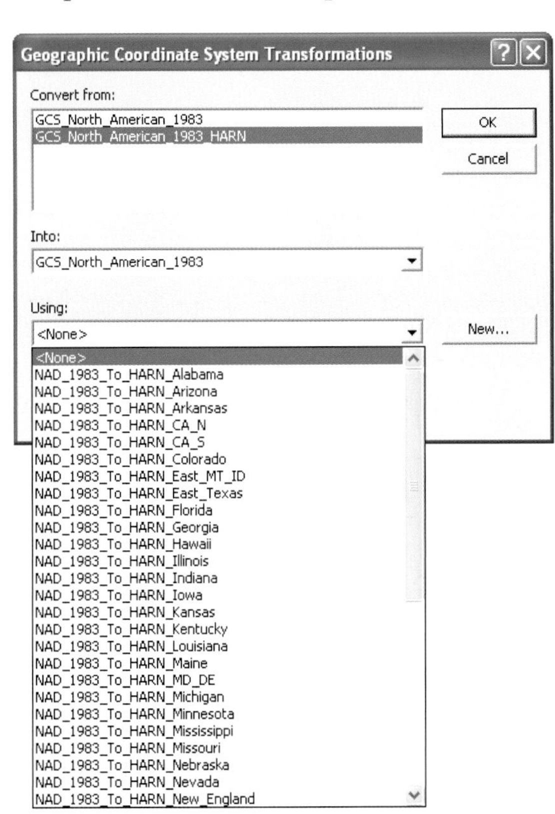

Figure 8-6 Even with only part of the list of available HARN transformations shown here, you can see it is extensive. The correct transformation must be selected from the drop-down list, in order to align the data correctly in the ArcMap data frame.

In this case the data is in Southern California, so the transformation NAD_1983_To_HARN_CA_S is selected to align the data. Zoom in on the data to see that the data is correctly aligned (figure 8-7).

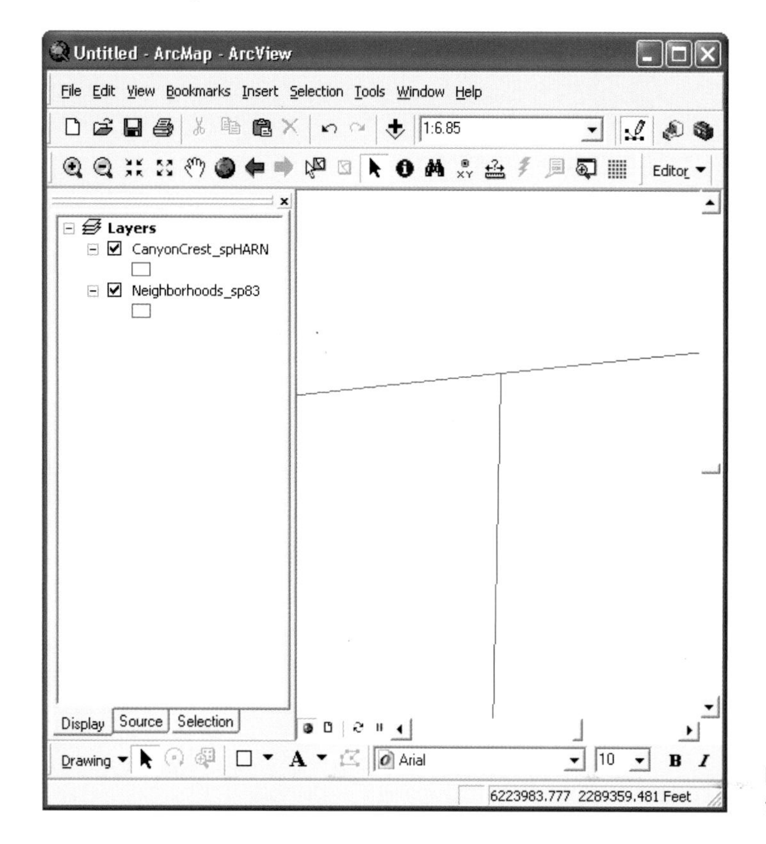

Figure 8-7 After applying the correct transformation, the data aligns properly.

If you remove the transformation from the ArcMap data frame, you can see the offset between NAD 1983 and HARN for this area of Riverside County in California. Figure 8-8 shows the offset.

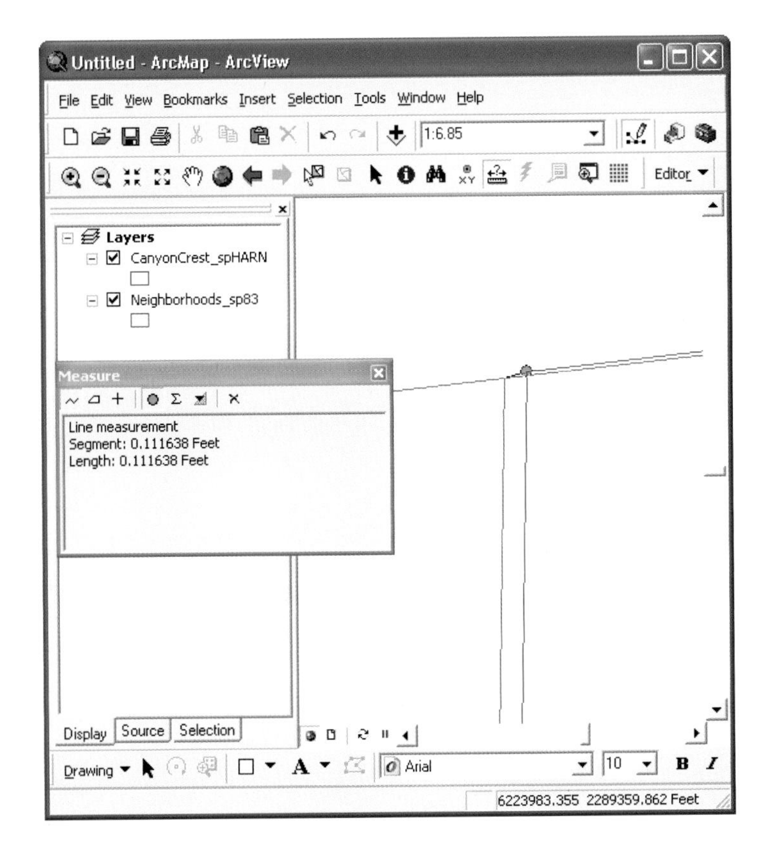

Figure 8-8 Offset between NAD 1983 (blue) and HARN (red) in Riverside County.

You are probably remembering that the offset between NAD 1983 and HARN is very small. However, if geodetic survey grade data is provided, which according to the National Geodetic Survey is supposed to be accurate to +/- 1 centimeter (about 4/10th of an inch or 0.0328 feet), setting the correct geographic transformation in the ArcMap data frame is still important.

Many geographic transformations create offsets that are much larger than that between NAD 1983 and HARN. Figure 8-9 shows the offset for these same data when the data are instead on GCS_North_American_1983 and GCS_North_American_1927.

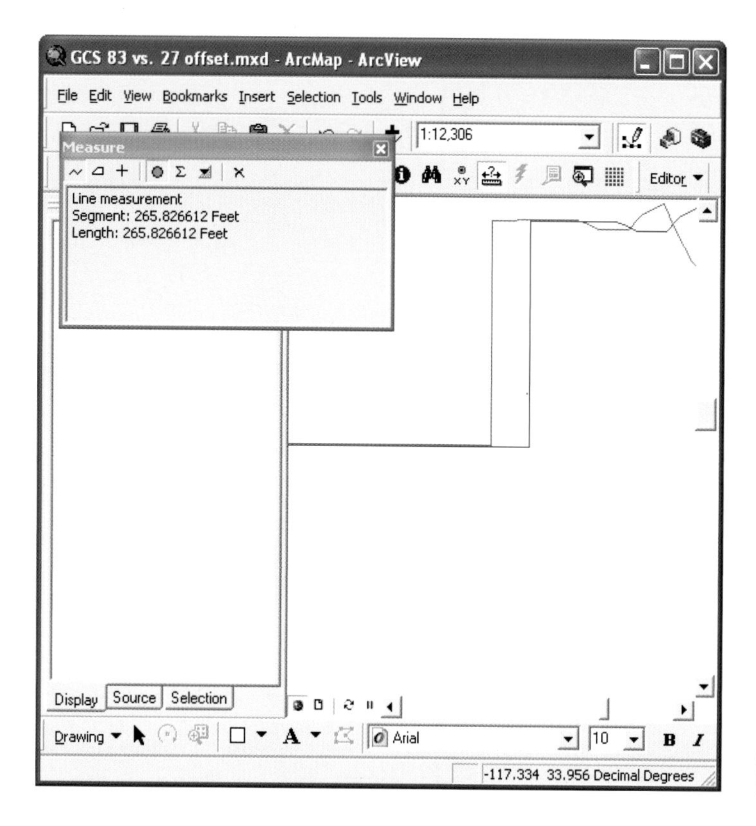

Figure 8-9 Offset between the NAD 1927 and NAD 1983 datums in Riverside County is hundreds of feet.

The measured offset between data on these two datums in this area is nearly 300 feet. In order to align this data in ArcMap, the correct geographic transformation must be applied to the ArcMap data frame. If the Geographic Coordinate Systems Warning dialog box is closed by mistake, the transformation can still be set in ArcMap: go to View > Data Frame Properties > Coordinate System tab, and click Transformations. The same Geographic Coordinate System Transformations dialog box is opened, and you can set the transformation for the area in which the data is located.

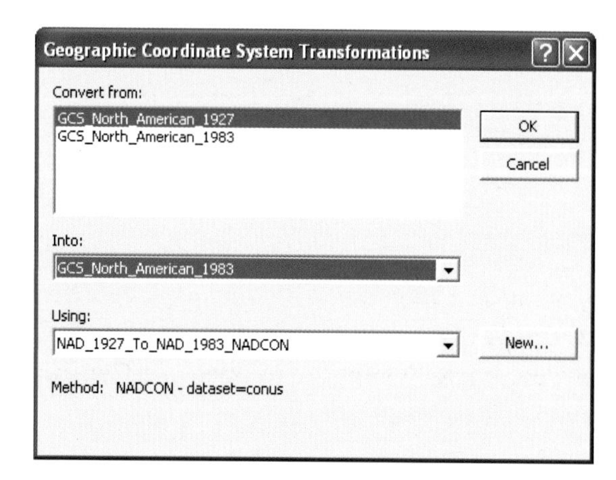

Figure 8-10 Setting the correct transformation between NAD 1927 and NAD 1983 for the contiguous forty-eight states. ("Conus" is an abbreviation for "contiguous United States" used by the National Geodetic Survey to designate the area of the United States, excluding Alaska and Hawaii.)

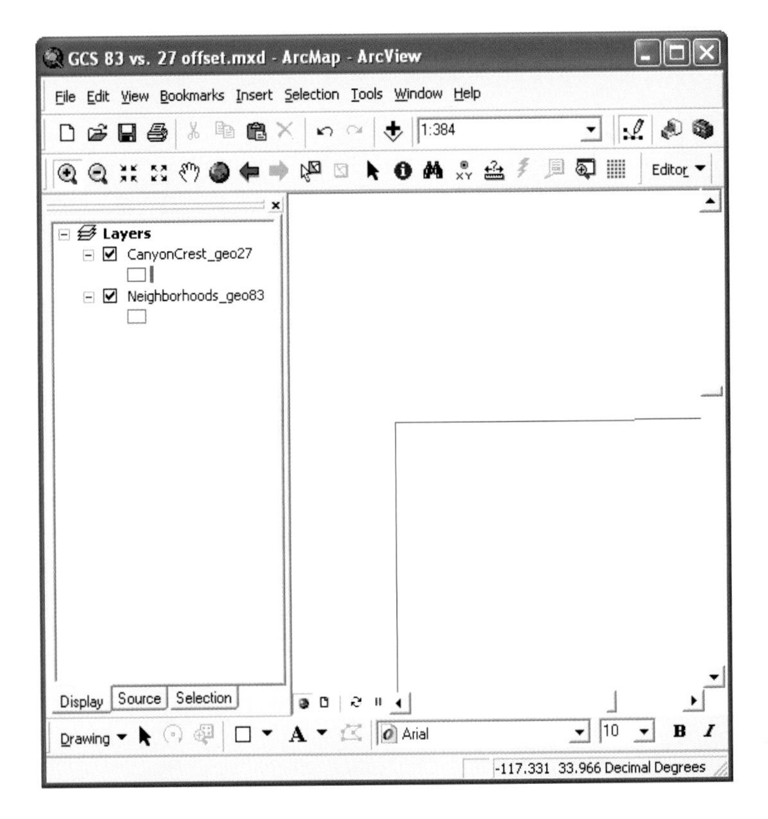

Figure 8-11 The blue line is directly underneath the red, which means that with the geographic transformation set correctly, the data is aligned.

CHANGING THE COORDINATE SYSTEM OF THE ARCMAP DATA FRAME

You have seen that when data with a defined projection is added to ArcMap, the coordinate system of the data frame is set automatically to the coordinate system of that data. However, the coordinate system of the data frame can be changed *during* the ArcMap session as well.

To change the coordinate system of the ArcMap data frame during the session, go to View > Data Frame Properties > Coordinate System, as shown in figure 8-2 (page 111). In the upper window labeled "Current coordinate system," the coordinate system definition for the data frame—with the parameters—is displayed.

In the lower window labeled "Select a coordinate system," you can open the Predefined folder, and access all the projection files installed with ArcGIS Desktop. Through the dialog box, you can select a different projection file, then by clicking Apply, the newly selected coordinate system will be applied to the ArcMap data frame.

If the new coordinate system has a different GCS from the data in the ArcMap document with which you are working, you will see a different geographic coordinate system warning dialog box displayed, as shown in figure 8-12. You must click Yes on this warning to dismiss this dialog box. Click the Transformations button on the right side of the Coordinate System tab to open the Geographic Coordinate System Transformations dialog box shown in figure 8-10.

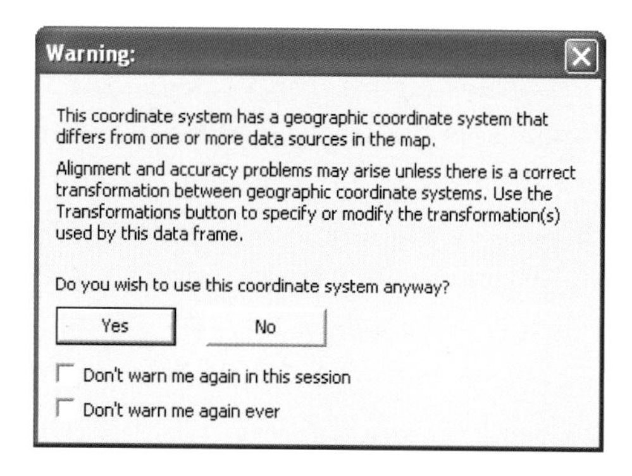

Figure 8-12 If the coordinate system of the ArcMap data frame is changed on the Data Frame Properties > Coordinate System tab, the warning shown here will be displayed.

Once you set the geographic transformation in the ArcMap data frame in this way, in the Geographic Coordinate System Transformations dialog box, the current ArcMap session will continue to use the selected transformation until you reset it. After setting the geographic transformation, you can check the box labeled "Don't warn me again in this session."

The box labeled "Don't warn me again ever" must *never be checked. This will permanently turn off geographic coordinate system warnings in ArcMap. As we have seen in previous chapters, these warnings are critical to maintaining proper data alignment in ArcMap.*

Only *direct* geographic transformations can be applied in the ArcMap data frame through the method as just described. A direct transformation converts between two GCS—for example, the transformation NAD_1927_To_NAD_1983_NADCON.

APPLYING GEOGRAPHIC TRANSFORMATIONS IN THE PROJECT TOOL IN ARCTOOLBOX

The Project tool is used *to create a new set of data* in a different coordinate system from the original data. This is something you would need to do if your goal is to perform comparative analysis between datasets, an operation on the data such as "Select by location," or editing in ArcMap.

The Project tool is located in ArcToolbox > Data Management Tools > Projections and Transformations > Feature. (To access this path in ArcGIS Desktop 10, from within ArcMap select Catalog window > Toolboxes > System Toolboxes.)

To use the Project tool and create output in a different projected or geographic coordinate system, the data selected as input must have the coordinate system correctly defined. If the input data selected does not line up properly with other data in ArcMap, *the input data has the wrong projection definition. Projecting the data to a different coordinate system will not fix the alignment problem.*

You need to verify that the coordinate system definition for the input data is correct before projecting the data, so before using the Project tool, open ArcMap with a new, empty map. Add the data you will be projecting to a new coordinate system, together with reference data for which you know the coordinate system is correctly defined. Once you have confirmed that the coordinate system for the input data is correctly defined, you can move on to create a new set of data. (If you find that the projection definition is not correct, refer to chapters 1, 2, and 3 for help fixing the projection definition.)

Open the Project tool, and select the input dataset or feature class to be projected. Select a suitable output folder for the new data, and assign a useful name to the new dataset that will be created. By default the Project tool will assign a name to the new dataset using the original name, but ending with _Project. Keeping the original name of the data is a good idea, but more helpful than adding _Project at the end of the name is to add information about the output coordinate system instead. Converting to a GCS, you might add _geo83 so that the dataset name tells you the data is in a GCS, with units of decimal degrees, on the NAD 1983 datum. When projecting to state plane, you might use _sp83 or _spHARN as the ending for the output dataset name, depending on the output coordinate system chosen.

In the Project tool dialog box, the last item is the geographic transformation list box. This is labeled "Geographic Transformation (optional)," as shown in figure 8-13.

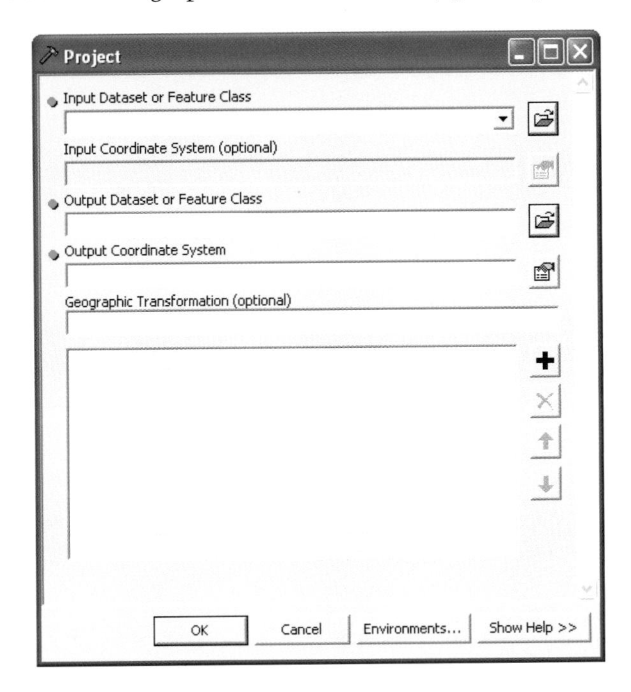

Figure 8-13 The Project tool dialog box.

The geographic transformation box is labeled as *optional* because if the GCS for the input and output coordinate systems are the same, no transformation is required. In figure 8-14, notice that the GCS for the input coordinate system is GCS_North_American_1983 (NAD 1983) and the output coordinate system GCS is also NAD 1983. The input and output data are on the same datum so no transformation is required. The "Geographic Transformation (optional)" box stays blank.

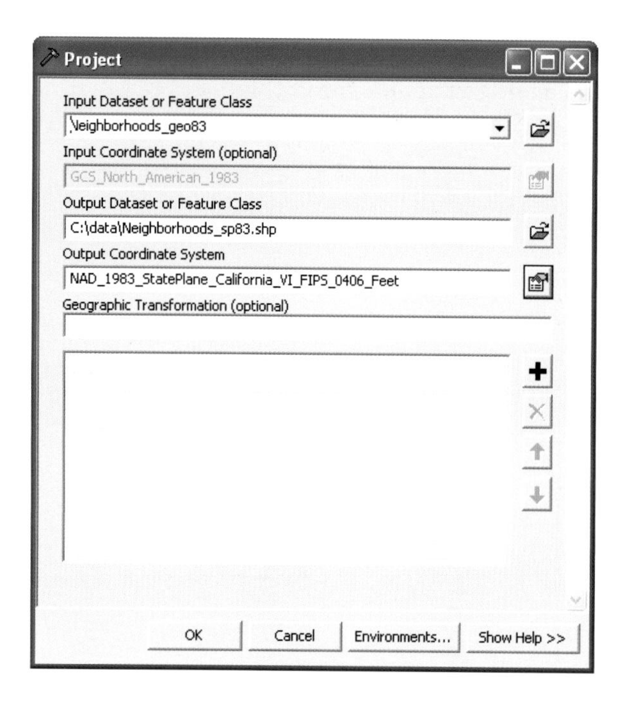

Figure 8-14 No geographic transformation is required because the input and output coordinate systems are both on the NAD 1983 datum.

If you select NAD_1927_StatePlane_California_VI_FIPS_0406 as the output coordinate system, the "Geographic Transformation (optional)" box turns into a drop-down, so that the required geographic transformation can be selected from the list, as shown in figure 8-15. If no transformation method is selected, the Project tool will fail with the error "Undefined Geographic Transformation."

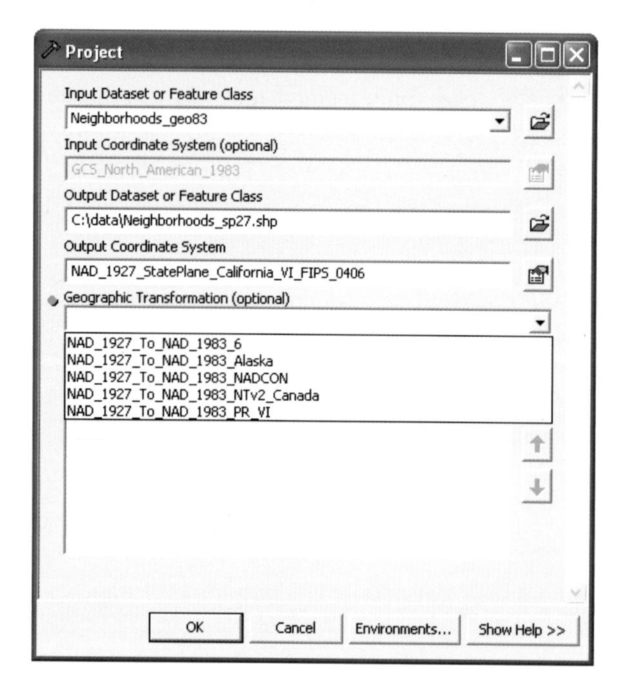

Figure 8-15 Because the data is in California, the transformation to select will be NAD_1927_To_NAD_1983_NADCON. Remember that the transformations work in either direction.

If you are projecting data that is on the NAD 1927 datum, and select HARN as the output GCS, you will have to select two transformations from the "Geographic Transformation (optional)" drop-down list. The first will be NAD_1927_to_NAD_1983_NADCON, and the second will be NAD_1983_To_HARN_xx for the state or area in which your data is located (figure 8-16).

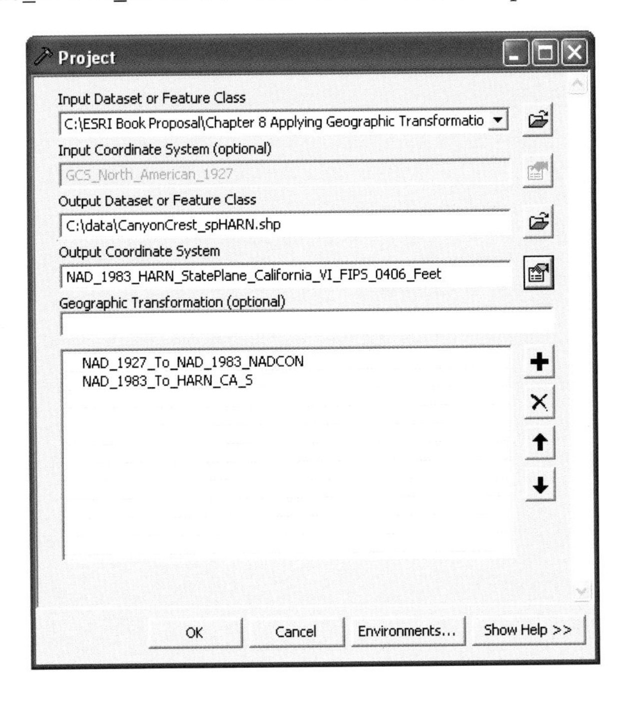

Figure 8-16 Specifying two geographic transformations in the Project tool. There are no direct transformations between the NAD 1927 and HARN datums. The Project tool will first apply the NADCON transformation, then apply the HARN transformation grid for this data in Southern California.

SUMMARY

This chapter examines the dialog boxes through which geographic transformations are applied in ArcMap, in order to align data displayed in the ArcMap data frame. Also reviewed is the Project tool dialog box, where new data in a new coordinate system and GCS can be created.

The next chapter describes types of projections and their properties. Chapter 9 also offers guidelines for important decisions about which projection is most appropriate to use for data maintenance and storage and for specific projects.

CHAPTER 9

"WHAT MAP PROJECTION SHOULD I USE FOR MY PROJECT?"

"What projection should I use for my data?"

There is no universally "correct" answer to the important question of what map projection you should use. There is simply the process of selecting what best serves your purposes. This chapter is intended to help you toward that end by examining in detail the differences between a geographic and projected coordinate system and the properties of various projected coordinate systems. The chapter suggests some criteria you can use in deciding which projection will provide the necessary results for your project.

GEOGRAPHIC COORDINATE SYSTEMS

A geographic coordinate system (GCS) uses angular units of measure; they are most often degrees. The first angle that indicates the location of data is given as longitude, the x-coordinate as measured from the prime meridian with its longitude value of 0. Any line of longitude can be used as a prime meridian in a GCS, but a mark at the Greenwich Naval Observatory in Greenwich, England, is most often used.

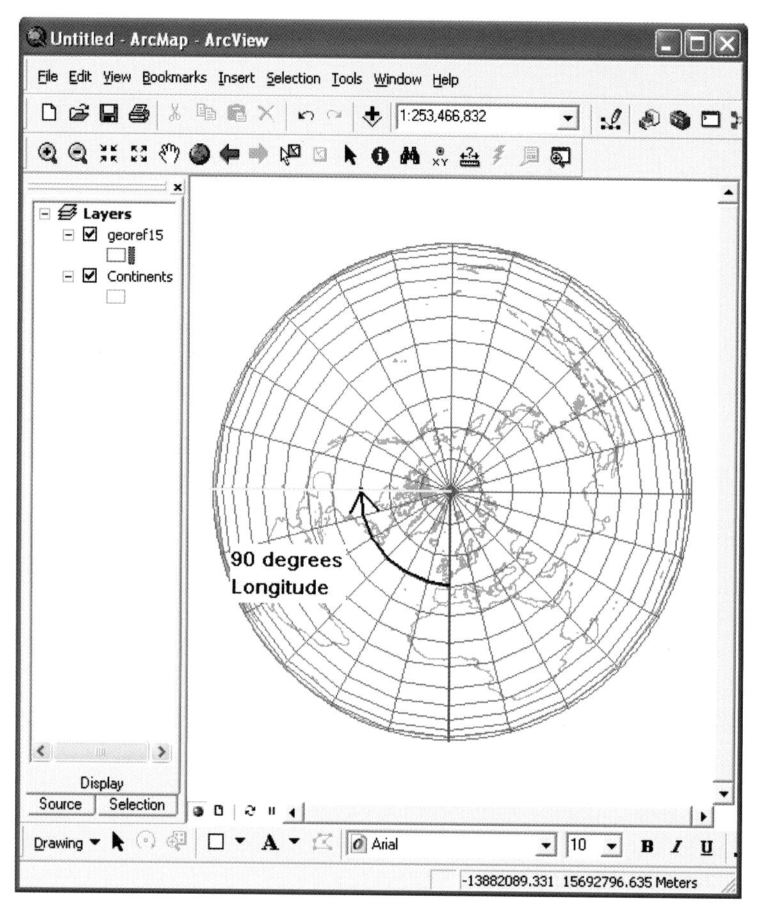

Figure 9–1 Lambert azimuthal equal area projection, showing the angular measurement of longitude, the x-coordinate, from Greenwich, England, to the point of interest. The yellow line indicates 90° west longitude, or -90°.

In figure 9–1 you are looking down on the earth from above the North Pole. The red line is the location of the prime meridian, which extends from the North Pole to the South Pole, passing through Greenwich, England. In our sample geographic coordinate system this is the prime meridian, assigned the value of 0 degrees of longitude. The yellow line represents the line of longitude that extends from the North Pole to the South Pole, passing through the point of interest that lies at 90° west longitude or -90°. The angle shown between the two lines of longitude is the value of the x-coordinate for the point of interest.

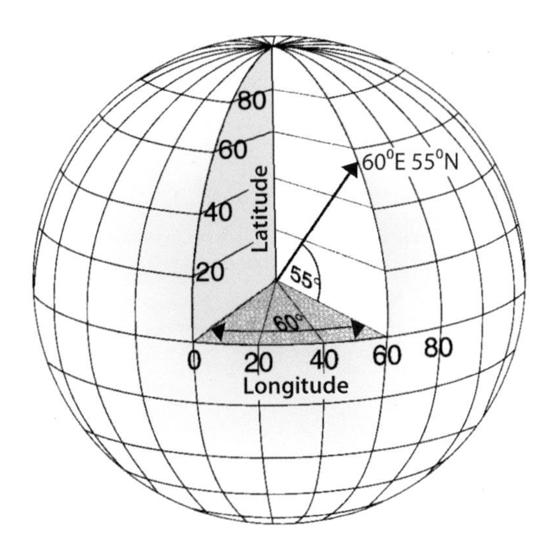

Figure 9-2 This is a 3D view of the earth, showing the east-west angle of longitude and the north-south angle of latitude.

In figure 9-2 you are viewing the earth in three dimensions. The angle of longitude in the easterly direction from the prime meridian is the x-coordinate for the point of interest, at 60° east longitude. The angle of latitude upward from the equator is the y-coordinate for the point of interest, at 55° north latitude. Note that the diagram in figure 9-2 does not specify a GCS.

LENGTH OF A DEGREE ON THE GROUND

There are 360 degrees in a circle. The linear distance that is represented by 1° along the outside edge of the circle depends on the size of the circle. A 1° angle is the same angle whether the circle has a radius of one meter or 6,378,137 meters, the latter of which is the radius of the earth at the equator as calculated for the WGS 1984 and GRS 1980 spheroids.

Spheroid definitions

Sources for WGS 1984 and GRS 1980 spheroid definitions are available at

`http://earth-info.nga.mil/GandG/publications/tr8350.2/wgs84fin.pdf`

An Album of Map Projections, by John P. Snyder and Philip M. Voxland. U.S. Geological Survey professional paper 1453, USGS: Washington D.C.,1989.

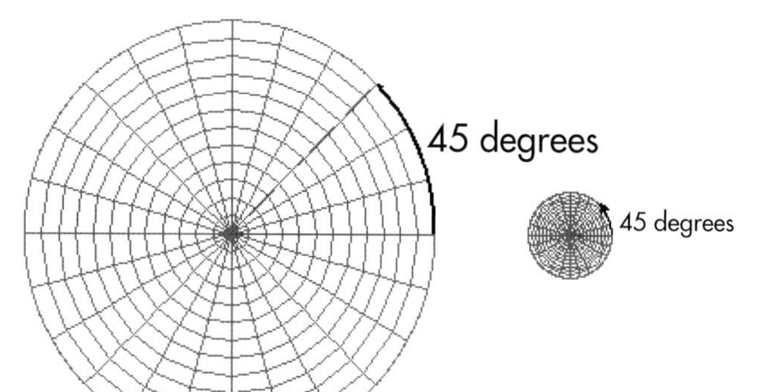

Figure 9–3 The angles in the diagram are both 45 degrees, but the linear distance around the outside of the circle, across the angle, is much longer on the large circle than on the small circle.

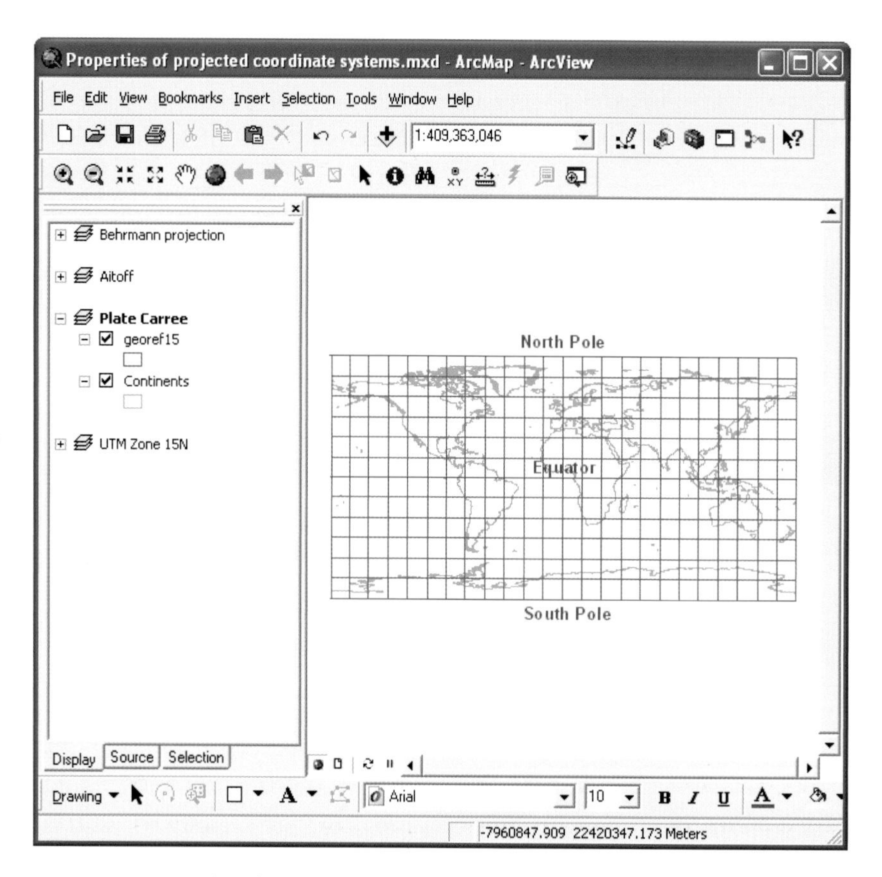

Figure 9–4 Displayed as a GCS, using the pseudo-plate carrée projection, the North Pole and South Pole, which are points, appear as long as the equator.

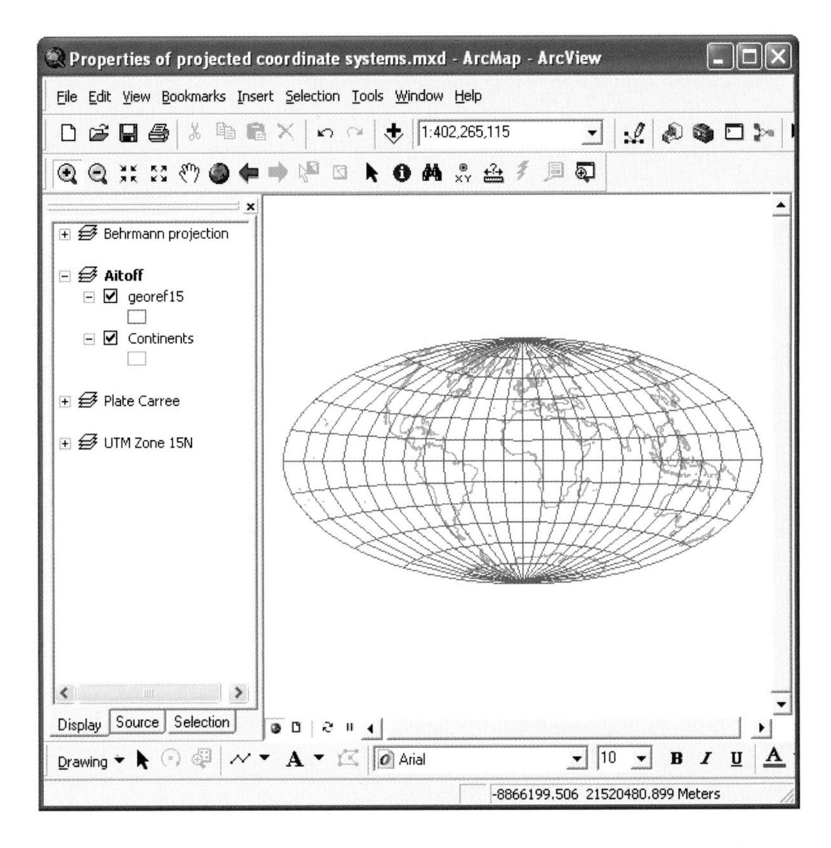

Figure 9-5 Displayed in a PCS, in this case the Aitoff (world) projection, the North Pole and South Pole appear correctly as points. Note how the lines of longitude converge toward the poles, and how the distance between the lines of longitude east to west decreases the farther north or south of the equator the latitude lies.

Lines of longitude, which wrap around the earth from the North to the South Pole and back again, and the equator are called great circles. These are lines of maximum distance around the earth. Along these lines around the earth, 1° represents a distance of approximately 111 km (69 miles) on the ground.

Lines of latitude are parallel to the equator. Because the earth is roughly spherical, the largest circumference is at the equator. Lines of latitude north or south of the equator represent shorter distances on the ground, depending on how far north or south of the equator those lines lie. At the North or South Poles, you have a point. The ground distance represented by 90° north latitude at the North Pole or -90° south latitude at the South Pole is zero.

Because the length of a degree is an angle, units of degrees cannot be used to measure distance on the ground between points, or the length of a line, because degrees of longitude and latitude represent different distances on the ground as the lines of longitude converge toward the Poles. An area of a polygon cannot be calculated in units of degrees because the angles have no relation to surface distances.

PROJECTED COORDINATE SYSTEMS

The problem of accurately measuring distances, lengths, or areas on the earth's surface led to the development of projected coordinate systems. Projected coordinate systems were needed so that distances and areas on the earth's surface could be measured in linear units like feet or meters.

LINEAR UNITS OF MEASURE

Linear units were needed for such diverse applications as navigation, taxation, and determining the area of a field to buy seed. Over the centuries, many different units have been used to measure distance and area, and different definitions of these units were proposed and applied, but eventually most of the world settled on the unit "meter." The standard definition of the meter today is based on the speed of light in a vacuum: exactly 299,792,458 meters per second. The meter is defined as the distance light travels in a vacuum in exactly 1/299,792,458 of a second. This converts to 39.37 U.S. inches. (See **`http://www.mel.nist.gov/div821/museum/timeline.htm`**)

There are many definitions of the unit "foot" used throughout the world. The unit known as the international foot, called "Foot" in ArcGIS Desktop, is equal to exactly 0.3048 of a meter. The U.S. survey foot, or "Foot_US," is equal to exactly 1,200/3,937 of a meter, or 0.304800609601219202438 40487680975... , which is an infinite, nonrepeating decimal. Because the international foot can be converted precisely to a meter when projecting data, some states have passed legislation that this unit of measure can or will be used with the state plane coordinate system for that state.

PROPERTIES OF PROJECTED COORDINATE SYSTEMS

Different types of projected coordinate systems were calculated to preserve different properties of data. The primary types of projected coordinate systems are:

- Conformal
- Equal area
- Equidistant
- True direction

In most cases, preserving one of these spatial properties of the data decreases accuracy of the other properties. In the case of small geographic areas such as a city or county, differences in distance or area calculated using these different types of projections for the same data can be very slight. When calculations are performed on data covering a large geographic area, differences in distance or area calculations can be substantial. The decision about what type of projection to use is critical when planning a project.

• Conformal

The most commonly used projected coordinate systems are conformal projections. These projections preserve the shape of the data but distort area, distance, and direction. A map of the earth using the Mercator projection in which Greenland appears as large as South America, and in which Antarctica is simply huge, is an extreme example of a conformal projection. Figure 9-6 displays the continents of the world with a 15° × 15° grid in this projection.

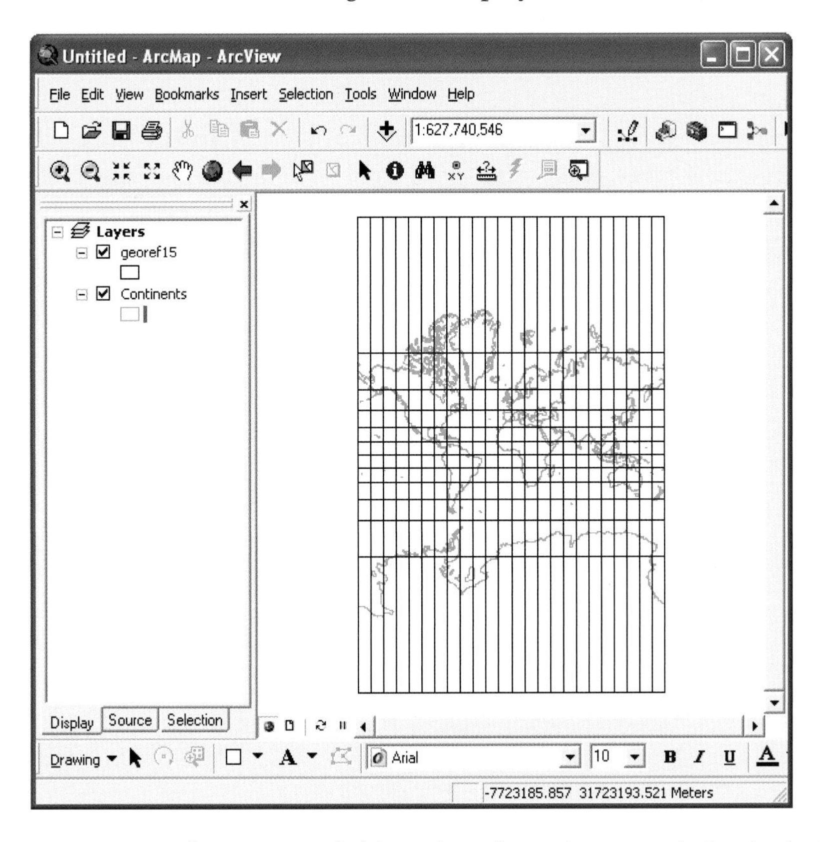

Figure 9-6 The 15° × 15° grid of the earth's surface and continents displayed in the Mercator projection, which is conformal. Compare the relative sizes of Greenland and South America and note the disproportionate extent of Antarctica in relation to the other continents.

- **Equal area**

In a case where accurate area measurements are most important, you should consider the use of an equal area projected coordinate system. The equal area property of this projection type is calculated so that the most accurate comparisons between the areas of polygons can be made. When utilizing an equal area projection for data, keep in mind that the shape, distance, and direction are distorted. The distortion in these other properties will increase as areas being examined become larger. Fortunately, some common uses for equal area projections, such as the land use comparison example for a city cited in chapter 8, affect such a small area that these distortions are not significant.

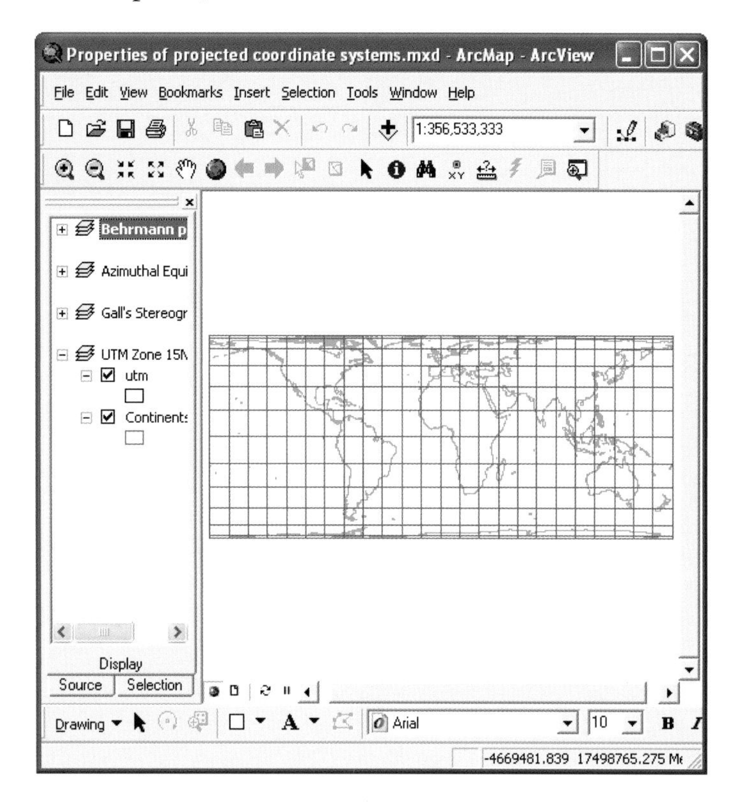

Figure 9-7 The continents of the world are shown with a 15°x15° geographic grid in the Behrmann projection, an equal area projection suitable for use in displaying data for the entire world. Notice that the black grid cells, which all measure 15° by 15°, are compressed in the north-south direction the farther the data lies from the equator.

- **Equidistant**

Calculating accurate distance measurements can be difficult. An equidistant projection will maintain only some distances. For instance, all north-south lines might have the correct length, or all distances calculated from the center point are accurate. Selection of an appropriate equidistant projection, therefore, should include testing various options and comparing the results from the tests with known measured or surveyed distances. The specific distortion will depend on the properties of the projection selected. Use of an equidistant projection will distort shape, area, and direction in the data.

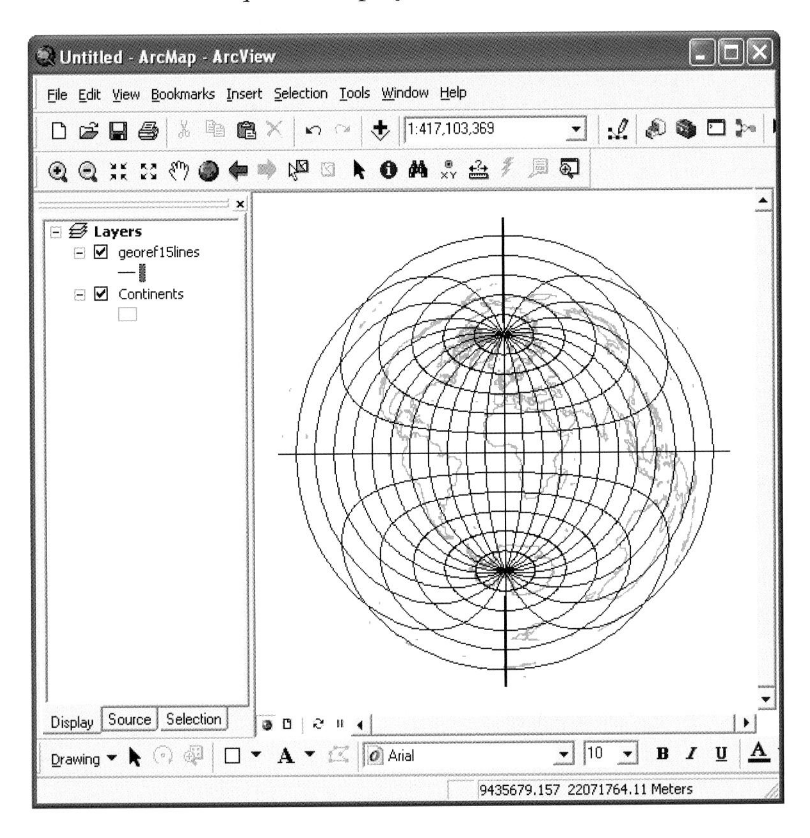

Figure 9–8 Displays continents and a 15° × 15° grid in the azimuthal equidistant projection, centered at the prime meridian and the equator. Distances measured from the center of the projection are most accurate. This projection also approximates true direction on a world scale from the center of the projection.

- **True direction**

A true direction projection is sometimes used for navigation. In a true direction projection, the angle between magnetic north and a line drawn on the map between the point of departure and the destination is the compass-bearing the captain will use to set the course for the ship or plane.

Further limitations exist for specific conformal, equal area, equidistant, and true direction projections. Some of these projections were calculated to preserve their respective properties over large geographic areas (small-scale map), while other projections will only preserve the desired properties over small areas (large-scale map).

In figure 9-9 the data is projected to universal transverse Mercator (UTM), zone 15 north (UTM zone 15N). UTM is a conformal projection that minimizes distortion in north-south trending areas. Each zone is 6° wide, except for some areas in Scandinavia. Use of the UTM coordinate system is limited to areas extending no more than 18° east to west. That will include the neighboring zones, in this case zone 14 to the west and zone 16 to the east. Beyond those limits, too much distortion is introduced into the data, so this coordinate system is not suitable for large east-west trending areas.

When the ArcMap data frame coordinate system is set to a specific UTM zone, as in this example, the display is limited to 45° east or west of the central meridian for the selected UTM zone. The central meridian for UTM Zone 15 is -87° west longitude, so the display at the western margin is cut off at -132° west longitude, while the eastern border of the display is at -42° west longitude.

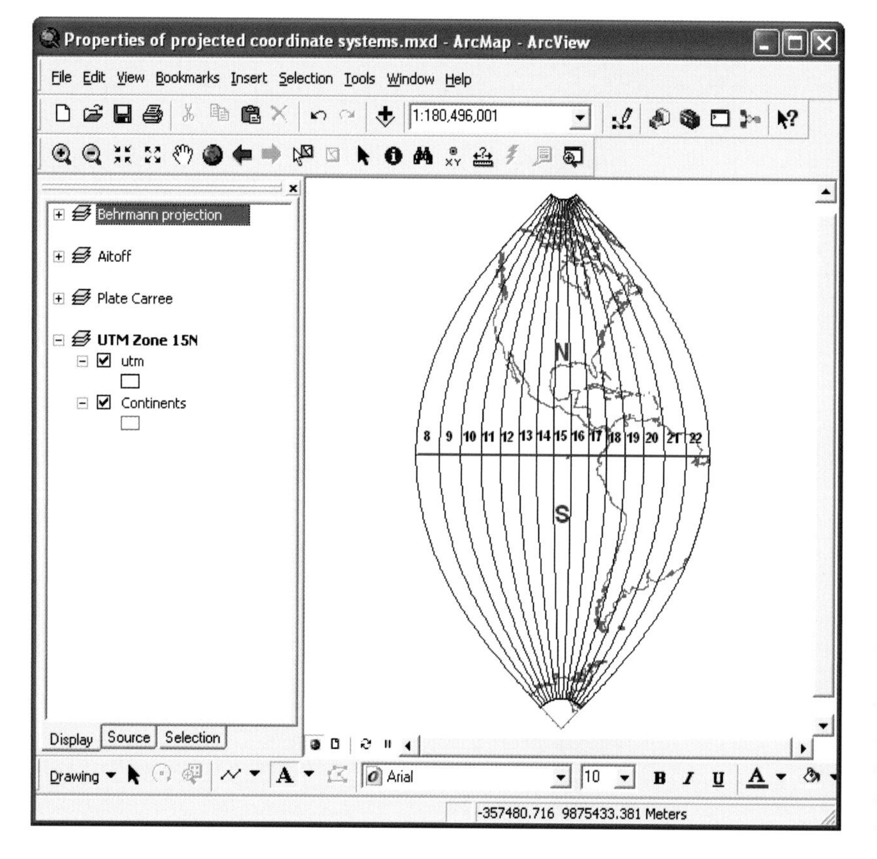

Figure 9-9 The UTM coordinate system, based on transverse Mercator, centered on UTM zone 15N. This image shows the display limits of the UTM coordinate system in the east-west direction as shown in ArcMap. The "N" and "S" denote zone numbers north or south of the equator, the red line in the image. To see the complete worldwide extent of the UTM coordinate system zones, refer to figure 3–7 (page 32).

DECIDING ON THE PROJECTION TO USE FOR YOUR PROJECT

You have been assigned a project for a client. The assignment is to make maps that allow you to compare historic land use for the county in 1979 with the present day. The client wants to know the areas of land in the county currently used for agriculture, residential, parks and recreation, industrial, roadways, schools and colleges, and public buildings such as government offices, police, and fire. What is the percentage change that has taken place in land use within the county over the past thirty years?

The *first* thing you need to do is decide on the map projection you will use for the project, in discussion with your boss and the client. In order to provide information so that the client can make an informed decision, you will need to project data for the task to different coordinate systems for comparison.

You cannot use a geographic coordinate system for the project because you are measuring and comparing *areas* and areas cannot be measured in decimal degrees. You must use a projected coordinate system with linear units such as feet or meters. Since more than fifty different projections are supported in ArcGIS Desktop, deciding on the PCS to use for the project appears complex.

Refer to appendix A and the Knowledge Base article 24646. The Projections Table referred to in the Related Information section of the article includes the names of supported map projections and information about the properties of each projection. Some projections are suitable for use in large-scale mapping—remember, large scale = small area—while other projections are suitable only for small-scale mapping—small scale = large area. In addition, some map projections are calculated for use within specific geographic regions of the globe. For example, Albers equal area conic, best applied in the midlatitudes and used for medium- to continental-scale mapping, is one projection that can be used within the forty-eight contiguous states.

Reviewing the table, you also see that these projections have different properties. Some preserve shape (conformal); others preserve distance (equidistant); others preserve direction (true direction); while still others preserve area (equal area). Under the "Properties" section of the table, look down the "Equal Area" column. You will see that there are a number of equal area projections: Albers equal area conic, Behrmann equal area cylindrical, Bonne, and Craster parabolic, to name just a few.

For this project, covering a small area, and comparing areas of land use, you want to select an equal area projection suitable for large-scale mapping. In the "Suitable Extent" section of the table, compare the areas of use for each projection checked in the Equal Area column under Properties. There is no equal area projection indicated as suitable for large-scale mapping, although Albers equal area conic indicates, "Distortion is moderate for most of the area" when this PCS is used for large-scale mapping. For comparison's sake, this will be a projection to try.

Here are some points to take into consideration in the initial planning stages of the project:

How accurate is your data? Your recent data could be extremely accurate, but how about the data from thirty years ago? How was that historic data created? Was the data digitized on-screen from older, scanned paper maps, a common method? Paper maps can stretch and be distorted from normal use; they are affected by humidity; and the scanning process itself can introduce distortion into the data.

Would using a standard coordinate system, such as state plane or UTM, provide results that are sufficiently accurate for your purposes? Remember that state plane and UTM, from the Supported Map Projections table (appendix A) are both conformal projections that preserve the shape of the data, but not direction, distance, or area. Over small areas, though, the distortion in these projections may not be large enough to affect the results.

In order to make a meaningful decision, you will need to do the following two things:

1. Overlay the historic data with the current data for the area of interest in ArcMap and compare the data. How good is the alignment of the data? Do major streets and tax parcels line up? If obvious distortions occur in the historic data, after allowing for rerouting of roads and other adjustments, it may be necessary to perform spatial adjustments on the historic data so that accurate comparisons can be made. It may also be necessary to create a custom projection file to align the historic data with current information, then project the older data and the current data to the same coordinate system.
2. Within the area of interest, obtain access to a surveyed area that is known to have an extremely accurate area measurement. For our purposes here, a fairly large area of 100 acres or more will be most useful. One acre is equal to 43,560 square feet (U.S. survey foot) or approximately 4,046.82 square meters.

Assuming that the historic and current data have reasonable alignment, project the current data to NAD 1983 StatePlane for the zone in which the data is located, or to NAD 1983 UTM for the correct zone. Units of feet or meters can be used. Compare the area of the polygon in that coordinate system with the area reported from the survey. Remember that both state plane and UTM are conformal projections, but over a small area the distortion in area is small. You may also need to compare the parcel boundaries in the current data with the survey, to ensure that the bearings and distances (metes and bounds) of the polygon correctly reflect the surveyor's work.

Calculating area for a shapefile in ArcMap

The attribute table of a polygon shapefile does not always include an area (SHAPE_AREA) field. Even if an area field exists, the values will not be recalculated when the data is projected to a different coordinate system. If the area field exists, you will need to recalculate the values after projecting the data to the new coordinate system. If an area field does not exist, you will have to add the field to the attribute table, and calculate the areas of the polygons manually.

Open the polygon shapefile attribute table, and check for an area field. If none exists, click Options > Add Field.

The field name must not exceed ten characters, must not contain spaces, cannot begin with a number, and cannot include special characters. It is useful to include an abbreviation for the units of measure in the field name, and assign names like AREA_usft or AREA_m.

Define the new area field as DOUBLE.

PRECISION is the total length of digits that can be stored in the field. Taking the default of 0 allows the maximum number of digits (19) to be stored in the field.

SCALE is the number of decimal places stored in the field. Taking the default of 0 allows the maximum number of decimal places (11) to be stored. Since the data is now in a projected coordinate system, using linear units, you may want to reset the SCALE value to a more realistic number such as 3.

To calculate the AREA, right click the field name, select Calculate Geometry, and select the units to be used for the calculation. The units need to match the units of the projected coordinate system.

Area, perimeter, and length are calculated automatically if data is stored in a file or personal geodatabase.

What is the difference between the areas? Is there a significant difference between the area calculated in the software and the surveyed area? Is the difference small enough that the analysis will not be affected, or is the difference large enough that the results of the analysis will be questionable?

If the data projected to standard state plane or UTM coordinate systems does not appear to match the surveyed extent of the test parcel closely enough, the next step will be to create a custom projection file for the area of interest, using the Albers equal area conic projection. Then you will project the current data to that custom coordinate system, and compare the area for your test parcel with the survey results as well as with the area in state plane or UTM. The client can then look at these samples and make an informed decision about which projection to use for the project.

CREATING THE CUSTOM PROJECTION FILE IN ARCMAP

Start ArcMap with a new, empty map, and add the current data that is projected to state plane or UTM.

To create the custom Albers equal area conic projection for your area of interest, you need to collect the following information to fill in the projection parameters.

Standard parallel 1

Standard parallel 2

Central meridian

Latitude of origin

Projection parameters

Projected coordinate systems have different parameters depending on the type of projection, extent of the area for which the projection is intended to be used, and other considerations. Here are some map projection parameters that may be required for various projected coordinate systems:

Standard parallel 1

Standard parallel 2

Central meridian or longitude of center

Latitude of origin or latitude of center

False easting and units of measure

False northing and units of measure

A complete list of supported projections and required parameters for each projection is available online at

http://webhelp.esri.com/arcgisdesktop/9.3/index.cfm?id=113&pid=112&topicname=List_of_supported_map_projections

With the current data displayed in ArcMap, go to View > Data Frame Properties > General tab, and change the Display Units to Decimal Degrees, then click Apply and OK. The display of units in the status bar at the bottom of the ArcMap Window will change from feet or meters to decimal degrees.

If the status bar is not displayed, click on View > Status Bar.

The central meridian for the projection is a line of longitude extending north to south through the approximate center of the data, dividing the data into an east half and a west half. Move your cursor across the data on the screen to a point that is at the approximate center of the data. It is convenient to round the central meridian value to an even decimal. For example, if the longitude value for the exact center of the data is 96.362758 you can round the value to -96.25 or -96.50. Remember that in the United States, you are west of Greenwich, England, so the longitude value for the central meridian will be a negative number.

Standard parallel 1 and standard parallel 2 are lines of latitude extending east to west across the data. Currently in ArcGIS Desktop standard parallel 1 is assigned the latitude closer to the equator, while standard parallel 2 is the line of latitude farther from the equator. However, in ArcGIS Desktop projection files, these can be reversed with no effect on data accuracy. (There is discussion about this issue in the GIS community. Be aware that some other software packages are unable to display data correctly if the standard parallel values are switched.)

To calculate standard parallel values for the Albers equal area conic projection, calculate the approximate north-south extent of the data, and divide by six. Standard parallel 1 will extend east to west, about one sixth the distance north of the southern edge of the data. Standard parallel 2 will be about one-sixth the distance south of the northern edge of the data. In most cases, you can judge suitable values for the two standard parallels by examining the data on screen. Rounding these values to two decimal places is also a good idea.

The latitude of origin can either be located at the center of the data or slightly south of the data's southern boundary. The advantage of locating the latitude of origin south of the data is that all y-coordinates for the data in this projection will be positive numbers. If the latitude of origin is in the center of the data, y-coordinates south of that line will be negative numbers. Again, this value can be rounded.

Write down the parameters and values you will use, then go to View > Data Frame Properties > Coordinate System tab. On the right side of the tab, click New > Projected Coordinate System.

In the New Projected Coordinate System dialog box type in a name for the new PCS. The name must not contain spaces or special characters, but underscores can be used. (Do not include the ".prj" extension here.) Select Albers from the Projection Name drop-down list, and enter the parameters you have collected in the Value columns.

ArcMap calculates the coordinate position of data to 16 significant digits in these fields so you will retain the string of 0s in the value that is located to the right of the decimal point. If your central meridian value is -96.5, the value in the field will be similar to the number shown below

-96.50000000000000

After entering the required parameters, select the units of measure you will use for the custom PCS. Select the same units of measure that you used for your original StatePlane or UTM projection, feet (U.S. or international) or meters.

For the Geographic Coordinate System, click Select. Select the same GCS that was used for your StatePlane or UTM projection. For comparison with data in the United States, open the North American folder.

If the original data is on:	Select the GCS named
GCS_North_American_1983	North American Datum 1983
GCS_North_American_1983_HARN	North American 1983 HARN
GCS_North_American_1927	North American Datum 1927
GCS_WGS_1984 (available with UTM, not State Plane), open the World folder	WGS 1984

On the New Projected Coordinate System dialog box, click Apply and OK. On the Coordinate System tab, click Apply. Watch the shape of the data change as the display is projected on the fly to the new projected coordinate system.

Click Add to Favorites to save a copy of the custom coordinate system to disk, then click OK. The custom coordinate system will be listed in the Favorites folder shown in the lower window on the Coordinate System tab.

When you click Add to Favorites, the custom projection file is saved to your User Profile location on your computer. See appendix C for the default location of your user profile, which depends on your version of ArcGIS Desktop and the operating system of your computer.

To make the new custom projection file easily available for defining coordinate systems and projecting data in ArcToolbox, open Windows Explorer, navigate to your user profile location, and copy the file. Navigate to the installation location for ArcGIS Desktop on your computer (refer to appendix B) and in the Coordinate Systems folder create a new folder. Name the new folder Custom PRJ Files or some other name that is meaningful to you. Paste the custom projection file into the new folder. This makes the custom projection file more accessible in ArcGIS Desktop and also ensures against loss of the file. Custom projection files stored in a custom folder you create are not removed when the software is uninstalled.

To quickly and easily project your data to the new coordinate system in ArcMap, right-click the name of the layer > Data > Export Data. In the Export Data dialog box, change the options button to use the coordinate system of the data frame; select a suitable output location on the local hard drive; give the output file a meaningful name; and click OK.

Add the exported data to the map as a layer. If the data was exported to a shapefile format, open the attribute table and recalculate the Area field to update the areas for the polygons in the new PCS. If the data was exported to an existing personal or file geodatabase, the area was recalculated automatically.

Open the attribute tables for both the original data projected to StatePlane or UTM, and the data in the new PCS, and compare the areas for your large parcel. Which area most closely matches the area from the survey? The area in the new Albers equal area conic projection? Or the area in the original PCS?

Now you have the information to present to the client, so that the client can decide which coordinate system to use for the project.

ACCURATE LENGTH MEASUREMENTS

This time, you have been assigned the task of accurately mapping the length of a pipeline. The pipeline extends across part of two states, and the data is currently stored in geographic coordinates with units of decimal degrees. You know the measured length of the pipeline from the construction records, but the length cannot be calculated in the attribute table in decimal degrees.

For this project, you cannot use the state plane coordinate system because each state plane projection definition applies to only a single area within each state, and in this case, the data crosses state lines. The pipeline also crosses UTM zone boundaries. Since transverse Mercator, the base projection for UTM, is a conformal projection, UTM would not be the best coordinate system to select either.

To most accurately calculate *length*, you will need a custom equidistant projection.

Note that lengths calculated from data in an equidistant projection are strictly 2D, sometimes referred to as horizontal or planar distance. Additional length of the linear features due to elevation changes will not be calculated in the projection process.

Refer to the Supported Map Projections table (appendix A, Knowledge Base article 24646) to locate an equidistant projection suitable for either regional- or medium-scale mapping. You will find there is no projection with both these properties, but both azimuthal equidistant and equidistant conic are suitable for regional-scale mapping, while equirectangular (equidistant cylindrical) can be used for large-scale mapping.

Referring to the link for supported map projections, you see that azimuthal equidistant requires the following parameter values:

Central meridian
Latitude of origin

Equidistant conic requires:
Standard parallel 1
Standard parallel 2
Central meridian
Latitude of origin

According to the table, azimuthal equidistant is most suitable for data in the equatorial region, while equidistant conic works best for data oriented east-west in the midlatitudes. Based on the location of the data in the United States, equidistant conic appears to be the better choice for a custom projected coordinate system.

Again, the central meridian is a line of longitude running north-south in the approximate center of the data, divides the data into an east half and a west half.

The latitude of origin runs east-west, and can be either south of the data or can cross the center of the data. The standard parallel values selected, though, divide the data into thirds, unlike the Albers projection that uses the one-sixth rule.

Change the Display Units to Decimal Degrees and collect the values you will enter for these parameters.

Go to View > Data Frame Properties > Coordinate System tab. On the right, click on New > Projected Coordinate System.

Enter a name for the new projection. Remember to use underscores and to avoid special characters in the name. Select Equidistant Conic from the drop-down list, and enter the parameter values you collected. Select the units to be used and assign an appropriate GCS. After applying the custom PCS to the ArcMap data frame, remember to click Add to Favorites, saving the new projection file to disk.

Export the current data to the coordinate system of the data frame. If the data is in a shapefile, open the Attribute table, and recalculate the Length or Shape_Length field if it exists. If no Length field exists, add the field as Type: Double, accept the default Precision of 0, and enter a Scale of 3.

Select the Length field, and using the Field Calculator, calculate the new Length values for the records in the table.

You can now compare the calculated length of the pipeline from your reprojected data to the actual measurements from the construction records.

WHAT ARE FALSE EASTING AND FALSE NORTHING?

You have noticed that false easting and false northing parameters are included for the Albers equal area conic and equidistant conic projections. Values in the units of the projected coordinate system can be entered for these parameters to make all values across the extent of the data positive numbers. This makes distance and area calculations more efficient, and can be important if data is exported to another software program. Here is how this works.

Refer to figure 1-10 on page 9; it shows the distribution of positive and negative coordinates for data in a geographic coordinate system across the world. In a geographic coordinate system, any x-coordinates that are west of the prime meridian are negative numbers. Y-coordinates south of the equator are negative numbers.

When a projected coordinate system is created, the central meridian serves as a sort of "prime meridian" for the new PCS, so any x-coordinates for data that are west of the central meridian will be negative.

The latitude of origin serves as the "equator" for the new PCS, so any y-coordinates south of the latitude of origin will be negative numbers.

You can see this in ArcMap by displaying your data in the new Projected Coordinate System dialog box, then moving your cursor across the screen from left to right or top to bottom. When you cross the Central Meridian or Latitude of Origin, you will see the x- or y-coordinates change from positive to negative values on the status bar. The central meridian and latitude of origin have become the lines of zero x and y for the new projected coordinate system.

People often prefer that the coordinate locations shown by the cursor on the screen are all positive numbers across the entire extent of their data, rather than displaying as a mix of positive and negative values. In order to make all the values positive, values are entered into the False Easting and False Northing field in the projection file.

To find an appropriate value for the false easting, add data to ArcMap, and set the coordinate system of the data frame to your custom PCS. Move your cursor across the screen in the east-west direction to find the point where the x changes from positive to negative numbers. Use the Measure tool, set the Distance units to match the units of your custom projected coordinate system, then measure from the central meridian (x = 0) west, to the left, past the western edge of your data.

Make a note of the distance measurement, and round the number up to an even value. For example, if the measurement is 269,541 feet, round up to 300,000 feet. This value is entered as the false easting.

If the y-coordinate value used for the latitude of origin is south of the data, all data in the area of interest will have positive y values. If the latitude of origin selected for the custom PCS is in the center of the data, y-coordinates south of this latitude will be negative.

If you wish, measure from the latitude of origin south past the southern extent of the data. Round this value up to an even number, and enter the value in the False Northing field.

Remember to save the modification to the projection file by clicking Add to Favorites, and copy the modified projection file into your Custom PRJ Files folder at the location shown for your version of ArcGIS and your operating system, as listed in appendix B.

Many other projections, with different properties, are available in ArcGIS Desktop. These different projections can be used for other projects with different objectives.

A clue of value

The state plane coordinate system projection files *all* include a false easting value. Examine the projection files in ArcCatalog at Coordinate Systems > Projected Coordinate Systems > State Plane. You will see a wide variety of false easting values in each projection file, entered to make all longitude (x) values across the extent of the state plane zone positive numbers. Some state plane projection files also include a false northing value to make all y-coordinates positive numbers within the zone, but this is less common. The same zone based on different GCS (NAD 1927 and NAD 1983) often will have different false easting or false northing values. This makes it possible to identify by inspection which GCS is being used.

UTM zone definitions across the entire world also include a false easting value. Since the UTM zones are all the same width—6°—the false easting is the same for all zones: 500,000 meters. In the UTM coordinate system, the latitude of origin is always 0° at the equator. This would make all latitude values south of the equator negative numbers. To avoid this, UTM zones south of the equator all include a 10,000,000 meter false northing value.

Adding a false easting or false northing value to the projection file has no effect on the accuracy of the projected coordinate system.

SUMMARY

Chapter 9 examines the properties of various map projections that are supported in ArcGIS Desktop. This chapter also offers examples of the way this information can be used to decide which projection will work best to extract necessary data for a specific project. These principles can also be applied when deciding which projection to use for maintaining or storing data.

The next chapter discusses the parameters included in a projection file, how values for those parameters are determined, and what purpose each serves. Chapter 10 also includes detailed information about adding x,y data to ArcMap, identifying the coordinate system of the data in the table, then converting that data to a shapefile or geodatabase feature class. The chapter also addresses the frequently asked question about the shape of buffers displayed in the ArcMap data frame.

CHAPTER 10

PROJECTION FILES AND PARAMETERS; ADDING X,Y DATA; THE SHAPE OF BUFFERS

"What do all these items in a projection file mean, and what do they do to the data?"

"When I add x,y data to ArcMap, the points draw in the wrong place. What do I do now?"

"Why aren't my buffers round in ArcMap?"

This chapter addresses additional questions frequently asked by GIS users. The unifying thread among the topics is again the matter of assigning correct projection parameters to data, so that the data display or analysis properly presents the needed information. Therefore, this chapter describes in greater detail the parameters of map projection files. A parameter is one of the variables that defines a particular map projection or coordinate system. Let's put this into context and be very clear why parameters are important to lining up data in ArcMap: All data is created in some coordinate system. When you identify which coordinate system your data is in, you can *define the data* correctly — *set or confirm the correct parameter values* — and it will draw in the right location in ArcMap in relation to other data. Basically, "defining" the data means selecting the map projection file (coordinate system) that correctly describes the coordinates for the data.

Through understanding these basic concepts, the answers to some frequently asked questions — about x,y data and about the shape of buffers, to name two — begin to make sense. For example, when adding x,y data from a table to ArcMap, the coordinates must be examined to determine the correct type of coordinate system for the data. Then the projection can be correctly defined while adding the data to the map. Regarding buffers, it is important to know that the shape of buffers for point features is controlled by the coordinate system of the input data and by the projection set for the ArcMap data frame.

PROJECTION FILES INSTALLED WITH ARCGIS DESKTOP

Refer to appendix B for the default installation location of ArcGIS Desktop. You will find in the installation directory a folder named Coordinate Systems, which includes upwards of seventy folders containing more than 4,000 projection files—standard coordinate systems in which data can be created. The sheer number of these makes it impossible for the user to randomly select a projection file to define the coordinate system for data and expect that particular projection file to line up the data in ArcMap. A systematic method for identifying the coordinate system of the data is essential.

The key to this process is the extent of the data. The data extent can be viewed and analyzed when the data is added to ArcMap. Data can be created in geographic, projected, or local coordinate systems. By examining the extent of the data in ArcMap, you can identify the type of coordinate system used to create the data. Then you can apply additional techniques to determine the precise coordinate system of the data. In special cases, you can also create custom projection files to align data, a process detailed in chapter 9.

COORDINATE SYSTEM PARAMETERS

The following parameters are required for all coordinate systems:

- Name
- Units of measure
- Datum

In geographic coordinate systems the units are angles. The most commonly used units are degrees; 360 degrees in a circle. Other angular units that can be used in projection files are radians, grads, gons, and microradians.

Projected coordinate systems use linear units such as meters or feet, and may require some of the following parameters in addition to the name, units, and datum: zone number; for UTM, Gauss-Kruger, some national grids, and state plane (United States and territories only).

Other parameters may be required for some projected coordinate systems and values may be provided by the data source in units of decimal degrees. **Decimal degree units are used in ArcGIS Desktop to define the following parameters for the coordinate system:**

- Central meridian or longitude of origin
- Standard parallel 1
- Standard parallel 2
- Latitude of origin
- Longitude of natural origin
- Latitude of natural origin
- Longitude of second point
- Latitude of second point
- Azimuth
- Rotation angle

False easting and false northing parameters are provided in linear units. The values will usually be in feet or meters, and will be the same as the units of the coordinate system.

Converting to decimal degrees

Projection files in ArcGIS Desktop only use units of decimal degrees (DD) for parameters. If projection parameters are given in Degrees–Minutes–Seconds (DMS), here is how to convert to DD using the Windows Calculator:

- Open the calculator at Start > Programs > Accessories
- Click View > Scientific
- Click View again, and make sure Decimal and Degree options are selected.
- Enter the Degree value followed by a decimal point, then the Minutes and Seconds values after the decimal.
- For example, if the DMS value is -115° 42' 23.75"
- Enter -115.422375 (note that only one decimal point is entered).
- Check the box labeled 'Inv'
- Click the button labeled 'dms'
- The output in decimal degrees is -115.706597222222222222222222222

Another example:

- DMS value is 43° 30' 48.8"
- Enter 43.30488
- Check the Inv box, then click dms
- DD value after conversion is 43.513555555555555555555555555556

WHAT IS A GEOGRAPHIC COORDINATE SYSTEM?

A geographic coordinate system (GCS) displays data within a grid of equal size cells, and uses angular units of measure (decimal degrees) to give coordinates for longitude and latitude. A geographic coordinate system has four components:

1. Spheroid (ellipsoid)
2. Datum
3. Prime meridian
4. Units of measure

WHAT IS A SPHEROID?

A spheroid (ellipsoid) is a mathematical representation of the shape of the earth calculated most recently from satellite measurements, although older spheroids calculated from ground measurements are also used in some cases. The terms spheroid and ellipsoid refer to the same mathematical model or shape, so to simplify the text, the term spheroid will be used.

The earth includes variations in elevation from the top of Mount Everest (at more than 29,000 feet) to the surface of the Dead Sea (at about 1,200 feet below sea level), plus the depths of the oceans. Methods for deriving a mathematical center of the earth and an average surface for the entire globe vary but they all result in a spheroid. Many spheroid values have been published over the years. These different spheroids can have comparatively large variations in size and shape.

A spheroid is a 3D shape created from a 2D model. The ellipse is an oval, with the major axis (the longer axis) through the earth at the equator, and the minor axis (the shorter axis) from the North Pole to the South Pole. A parabolic curve is drawn connecting the ends of the axes to create an ellipse. Rotate the ellipse around its minor axis and the shape of the rotated figure is a spheroid.

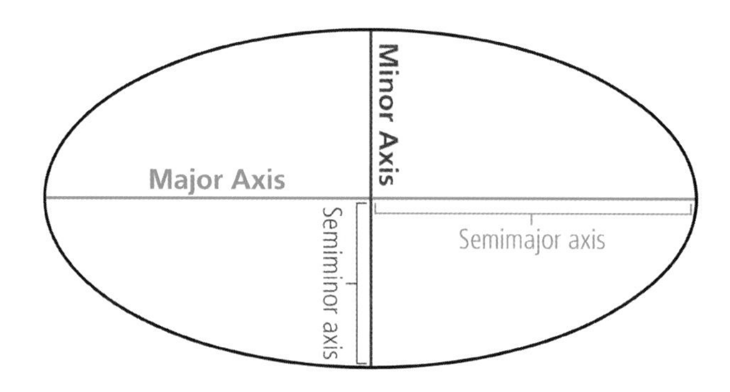

Figure 10-1 The major and minor axes of the ellipsoid and their position relative to the earth. The illustration shows how the semimajor and semiminor axes values for the spheroid are derived from the ellipsoid measurements.

The semimajor axis is half the length of the major axis. The semiminor axis is half the length of the minor axis.

One particular spheroid is distinguished from another by the lengths of the semimajor and semiminor axes and by the curvature of the surface defining the spheroid. For example, compare the Clarke 1866 spheroid with the GRS 1980 and the WGS 1984 spheroids, based on the measurements (in meters) below:

Spheroid	Semimajor axis	Semiminor axis
Clarke 1866	6378206.4 m	6356583.8 m
GRS 1980	6378137 m	6356752.31414 m
WGS 1984	6378137 m	6356752.31424518 m

A particular spheroid can be selected for use in a specific geographic area because that spheroid does an exceptionally good job of modeling the curvature of earth's surface *for that part of the world.* For North America the spheroid of choice is GRS 1980, on which the North American Datum 1983 (NAD 1983) is based.

WHAT IS A DATUM?

Datums—the collection of points of known accuracy used to georeference map data—can be horizontal or vertical. (A vertical datum provides a zero control point for elevations or depths and will not be discussed further in this book.) A horizontal datum is a reference for measuring longitude and latitude for a particular area of the earth's surface. A horizontal datum can be local, representing the average surface for an area as small as a single island; for example, St. Lawrence Island, part of Alaska. It can be a datum representing a country, such as the Japan Geodetic Datum 2000 (JGD 2000); a continent, such as the European Datum 1950 (ED 1950), the South American Datum 1969 (SAD 1969), or the North American Datum 1983 (NAD 1983); or the entire world, such as World Geodetic System 1984 (WGS 1984).

The underlying datum and spheroid to which coordinates for data are projected change the coordinate values. The following example uses coordinates for a location within the city of Bellingham, Washington, USA.

Compare the coordinates in decimal degrees for Bellingham in the NAD 1927, NAD 1983, and WGS 1984 datums. It becomes apparent that the coordinates expressed by the latter two of the datums are nearly the same, but the first one varies significantly: The coordinates of Bellingham on North American Datum 1983 (NAD 1983) and World Geodetic System 1984 (WGS 1984) are less than 2 meters apart for the same point, while the coordinates for the same place on North American Datum 1927 (NAD 1927) are quite different from the other two. This is because the underlying shape of the earth is expressed differently by the datum and spheroid. For comparison, the point on NAD 1927 is almost 100 meters or about 318 feet away from the point on the NAD 1983 datum.

Datum	X-coordinate Longitude	Y-coordinate Latitude
NAD_1927	-122.46690368652	48.7440490722656
NAD_1983	-122.46818353793	48.7438798543649
WGS_1984	-122.46819775227	48.7438850705687

The x-coordinate is the measurement of the angle from the prime meridian at Greenwich, England, to the center of the earth, then west to the longitude of Bellingham, Washington. The y-coordinate is the measurement of the angle formed from the equator to the center of the earth, then north to the latitude of Bellingham. (For an illustration of these angular measurements, refer to figures 9-1 and 9-2 on pages 124 and 125.)

WHAT IS A PRIME MERIDIAN?

A line of longitude extends north to south from the North Pole to the South Pole. In a geographic coordinate system, the prime meridian is the line of longitude where the x-coordinate is 0 (zero). Longitude values east of the prime meridian are positive numbers; longitude values west of the prime meridian are negative numbers (figure 1–10 on page 9 illustrates this).

The prime meridian most often used is the line of longitude, extending from the North to the South Pole, through the Greenwich Observatory in England. Other prime meridians are also used in geographic coordinate systems: Longitudes for points in Paris, Oslo, Beijing, and Jakarta are some of these alternative prime meridians. These alternative prime meridians center the data at that location, and in those geographic coordinate systems the alternative prime meridians will have a value of 0 degrees longitude.

WHAT IS A PROJECTED COORDINATE SYSTEM?

A projected coordinate system (PCS) is a type of coordinate system used for data if a precise location, distance measurements between features, lengths of linear features, or areas of polygons must be calculated. This is because a projected coordinate system is a 2D Cartesian system that simplifies calculations. A PCS is also used if the exact compass bearing between features must be measured.

WHY USE A PROJECTED COORDINATE SYSTEM INSTEAD OF A GEOGRAPHIC?

In a geographic coordinate system, locations are measured in angles from the known point of origin of the GCS. Because the earth is about 12,756,274 meters (12,756.274 kilometers or 7,926 miles) in diameter and 40,075,017 meters (40,075.017 kilometers or 24,901 miles) in circumference at the equator, 1° at the equator is about 111,319.5 meters (111.32 kilometers or 69.17 miles). Coordinates in decimal degrees do not provide very accurate placement of a point location, unless the value is carried out to many decimal places.

Because the units for a GCS are angles, the angle cannot tell you anything about distance on the ground. If you say a line is 1° long on the surface of a ball, it does not matter if the diameter of the ball is 1 meter or 12,756,274 meters. The length of the line is still 1°, even though on our one-meter ball 1° measures about 8.7 millimeters, while on the earth's equator the 1° angle measures 111,319 meters on the ground. The angular measurement is still only 1° and doesn't tell the user anything about distance on the surface. (For an illustration, refer to figure 9–3 on page 126.)

Area is length multiplied by width, and the length and width have to be measured in linear units. An area calculated in decimal degrees multiplied by decimal degrees gives an "area" in square decimal degrees or squared angles. Mathematically this makes no sense. That is why you sometimes need to use projected coordinate systems for your data. A GCS cannot be used when measuring distances or calculating areas because the length of a degree has no relation to surface distances.

Geographic coordinate system angles also introduce an additional complication. The equator circles the earth halfway between the North and South Poles. Being a circle, it is 360° around, and measures about 24,901 miles or about 40,075 kilometers. This is called a "great circle," a circle of maximum distance around the earth.

Any circle drawn around the earth, from the North Pole to the South Pole and back, is also a great circle and measures 360°. These circles are lines of longitude and the values provide x-coordinates for the position of data.

Say we have two lines of longitude that are 1° or 111,319 meters apart, measured east to west, at the equator. As the lines of longitude are drawn north toward the North Pole, the lines must converge because the North Pole is a point. Near the North Pole, then, the lines of longitude are still 1° apart but have almost no distance between them.

Since the convergence of the lines is continuous from the equator to the pole, 1° east to west at 45° latitude is 78,847 meters; at 80°, that degree is 19,393 meters; and at 89°, it's 1,949 meters. You can see from these numbers that coordinates in degrees cannot provide consistent distance measurements.

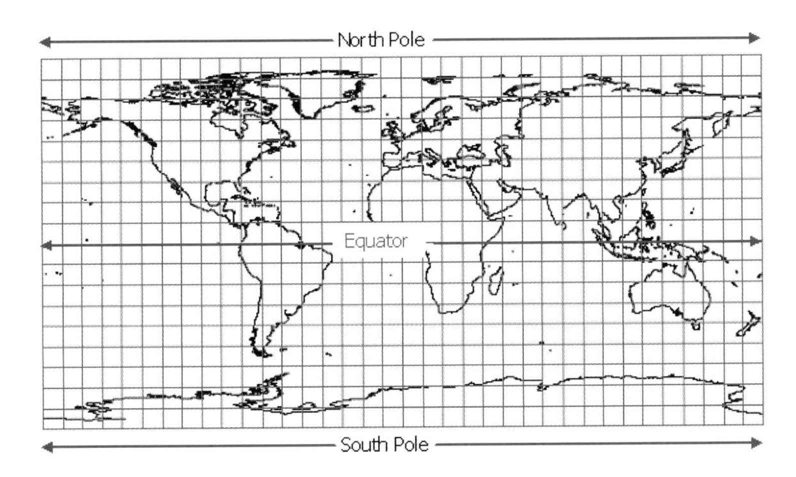

Figure 10-2 A 10°x10° grid covering the entire earth's surface, displayed over the continental areas.

Figure 10-2 represents a 10°x10° grid of the earth's surface. The data is being displayed in a pseudo-plate carrée projection, which treats the angles as if they are linear units. Because degrees are angles, each square is 10° on a side, even though the North Pole and South Pole, across the top and bottom of the grid, are supposed to be points.

PROPERTIES OF PROJECTED COORDINATE SYSTEMS

When projecting the rounded surface of the earth onto a flat piece of paper, different properties of the data will be preserved. When one of these properties is preserved in a particular projection, the other properties will be distorted. The **properties of map projections** are usually classified as

Shape

Equal area

Equidistant

True direction

Because each of these properties can be important for certain projects, many map projections have been created to preserve each of these specific properties for data: for certain purposes, for specific extents, for specific areas of the world, and for data that has a particular shape. These properties are mutually exclusive. You cannot have a map projection that preserves both the shape of data and distance. If the area being mapped is small (large scale) like a city, the distortion of the data may not be obvious, but the distortion will exist. If the area shown in the map is the entire world (small scale), the distortions inherent in different types of map projections will be very visible.

If preserving the **shape** of the data is most important, you would select a **conformal** projection. You have probably seen a map of the world where Greenland looks nearly as large as South America. Figure 10–2 illustrates an example. The shape of the data is preserved, but distance and area are very much distorted—Greenland is actually about one-tenth the size of South America.

If your project requires that **area** be measured most accurately, you would select an **equal area** projection for the data. Figure 9–7 illustrates an equal area projection, Behrmann, for the entire world. Other equal area projections are suitable for smaller geographic areas.

If your project requires that **distance** be measured most accurately, you would select an **equidistant** projection for the data, although not all distances can be preserved. Generally, all east-west or north-south distances are preserved, or all distances from the origin point of the projection. No equidistant projection can be calculated for the entire world.

If you are flying a plane or sailing a boat and need to **plot your course**, you would select a **true direction** projection so that the angle of the line drawn on the map would be the course you fly or sail to get to your destination. The azimuthal equidistant projection, illustrated in figure 9–8, approximates true direction on a world scale.

WHAT IS A LOCAL COORDINATE SYSTEM?

Local coordinate systems are very often used when creating computer-aided design (CAD) files. An arbitrary point is selected on the ground, often at a street intersection, property boundary corner, a survey monument, or other point. That point will become the 0,0 point for survey measurements. In the CAD program, bearings and distances for parcel data would then be drawn based on survey measurements from that point location.

ADDING X,Y DATA TO ARCMAP AND CONVERTING TO FEATURES IN A SHAPEFILE OR FEATURE CLASS

Longitude and latitude coordinates (x,y data) are frequently provided to GIS users in digital form. The coordinates may be stored in a spreadsheet or database table, in a space-delimited text file, or a comma-delimited text file. The coordinates in the file are most often in units of degrees, but may be in Degrees–Minutes–Seconds (DMS), rather than Decimal Degrees (DD). Coordinates in DMS will be formatted as shown in the example below:

```
-80 37 40     35 28 48
```

Coordinates provided in DMS need to be converted to DD before the points can be displayed in ArcMap.

Appendix A's Knowledge Base articles 22455 and 27548 from the ESRI Support Center contain VBA scripts that can be installed in ArcMap to perform this conversion. Instructions for installing the scripts are also included in the articles. Article 37264 contains script and instructions for ArcMap 10.

To convert these points to data in your GIS, you also need to make sure that the signs for the values are correct. The image in figure 1-10 on page 9 shows the distribution of positive and negative longitude and latitude values for locations across the earth. For x,y data in North America, the x-coordinates (longitude values) must be negative, while the y-coordinates (latitude values) are positive numbers.

Knowledge Base article 30727 provides additional instructions on formatting a table in Microsoft Excel so that the table and field names can be read by ArcMap.

ADDING THE X,Y DATA TABLE TO ARCMAP

There are two methods for adding the table containing x,y data to ArcMap. One good one is to click Add Data, navigate to the directory where the table is stored, and add the table directly to the ArcMap session (see figure 10-3).

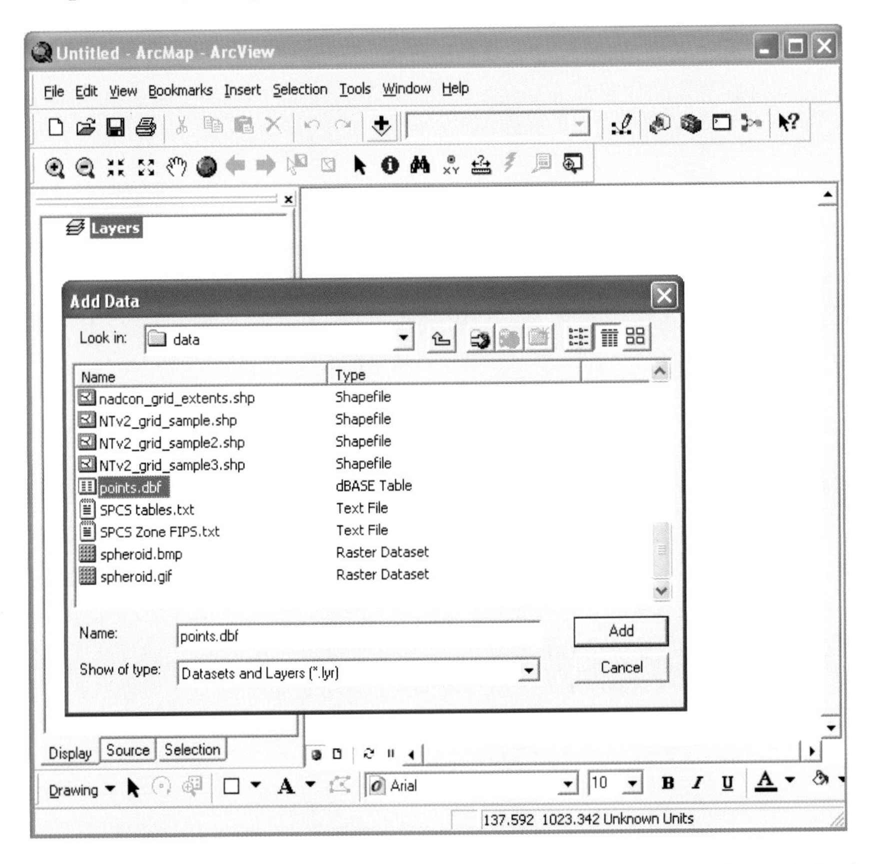

Figure 10-3 Select the table containing the point data through the Add Data dialog box in ArcMap and add the table to ArcMap.

As noted above, the table coordinates may be given in decimal degrees or in units of DMS. They might even be in feet or meters, using a projected coordinate system such as state plane or UTM. Adding the table to ArcMap, then opening the table—*before* defining the projection—will allow you to examine the table, find the names of the fields that contain the coordinate values, and decide on the most likely coordinate system definition for the values.

In this example the coordinates are in decimal degrees, as displayed in figure 10-4. The names of the fields containing the x- and y-coordinates are conveniently named POINT_X and POINT_Y.

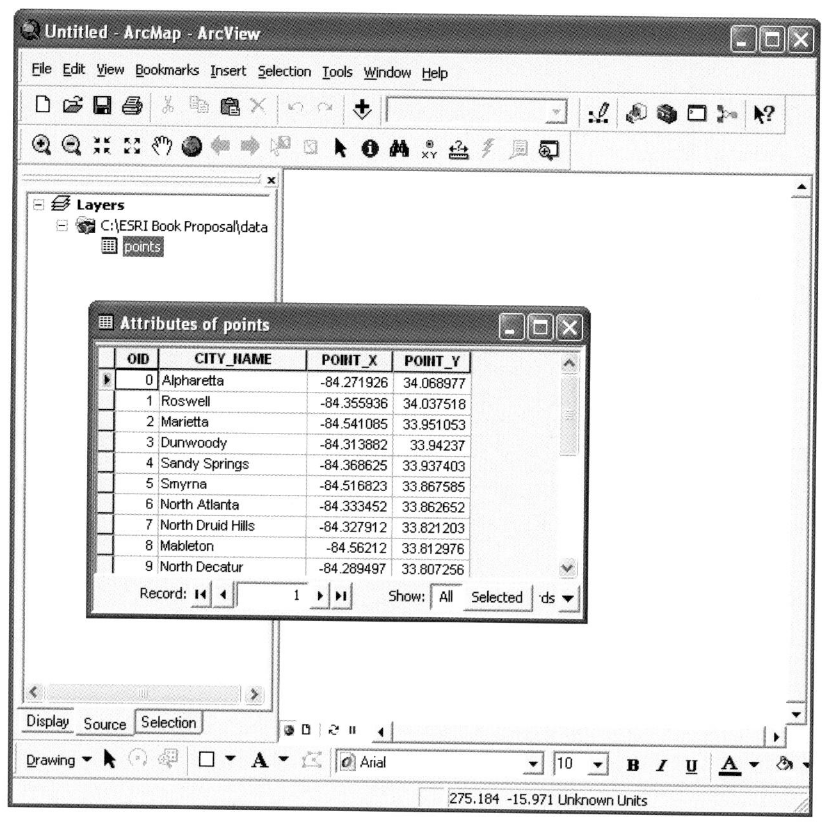

Figure 10-4 These coordinates are in decimal degrees and show points for cities in the United States.

To display the points in ArcMap, right-click the name of the table and select Display XY Data. In this dialog, populate the X Field and Y Field boxes with the field names that contain the x and y coordinates, as shown in figure 10-5.

ArcMap will populate the X Field and Y Field boxes automatically, *but the values selected for the field may be wrong*. This is fixed by clicking on the field drop-down list and selecting the correct field name that actually contains the proper coordinate values.

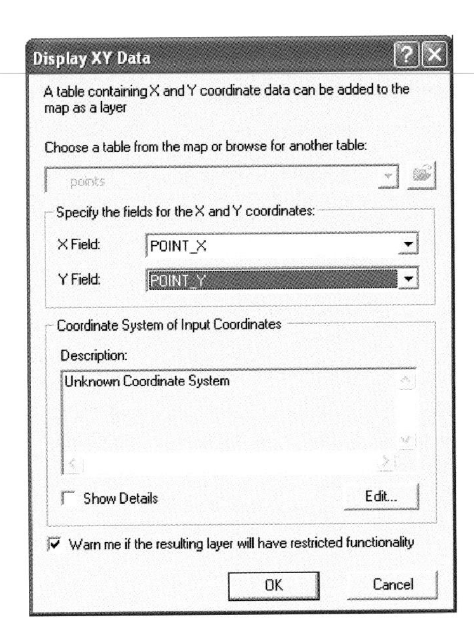

Figure 10-5 The fields containing the coordinates x (longitude) and y (latitude) are selected to populate the X Field and Y Field boxes in the dialog.

In order for the data to align with other data in the map, the projection must be defined for the data in the table. Click Edit to open the Spatial Reference Properties dialog box, then click Select. In this example, since the data has coordinates in decimal degrees, we will open the Geographic Coordinate Systems folder > North America. The data source informed us that the coordinates were on North American Datum 1983, so we will select that projection file from the available options. If we did not have that information from the data source, we would still begin by selecting this option for data within the United States. (Other techniques for identifying the geographic coordinate system for data are discussed in detail in chapter 2.)

Double-click the selected projection file. On the Spatial Reference Properties dialog box, click Apply and OK, as shown in figure 10-6.

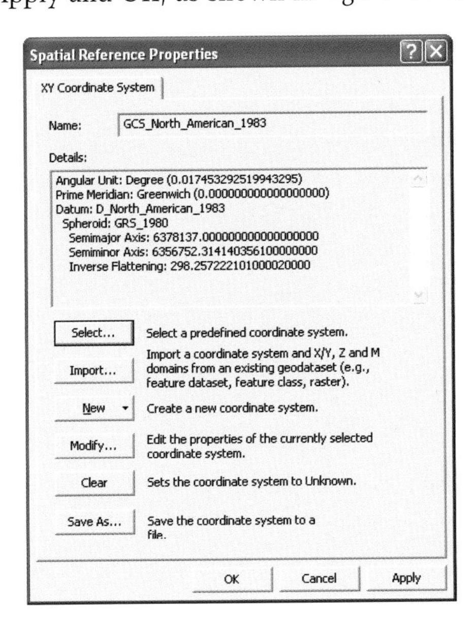

Figure 10-6 The correct coordinate system definition has been selected for the point data. Click Apply and OK on the Spatial Reference Properties dialog box.

Click OK on the Display XY Data dialog box.

The point data will display in ArcMap, listed as an Events theme in the ArcMap table of contents. In relation to the shapefile "usstpln83.shp" in figure 10-7, you can see that these points display in the Georgia West StatePlane FIPS zone. This is the correct location for the point data.

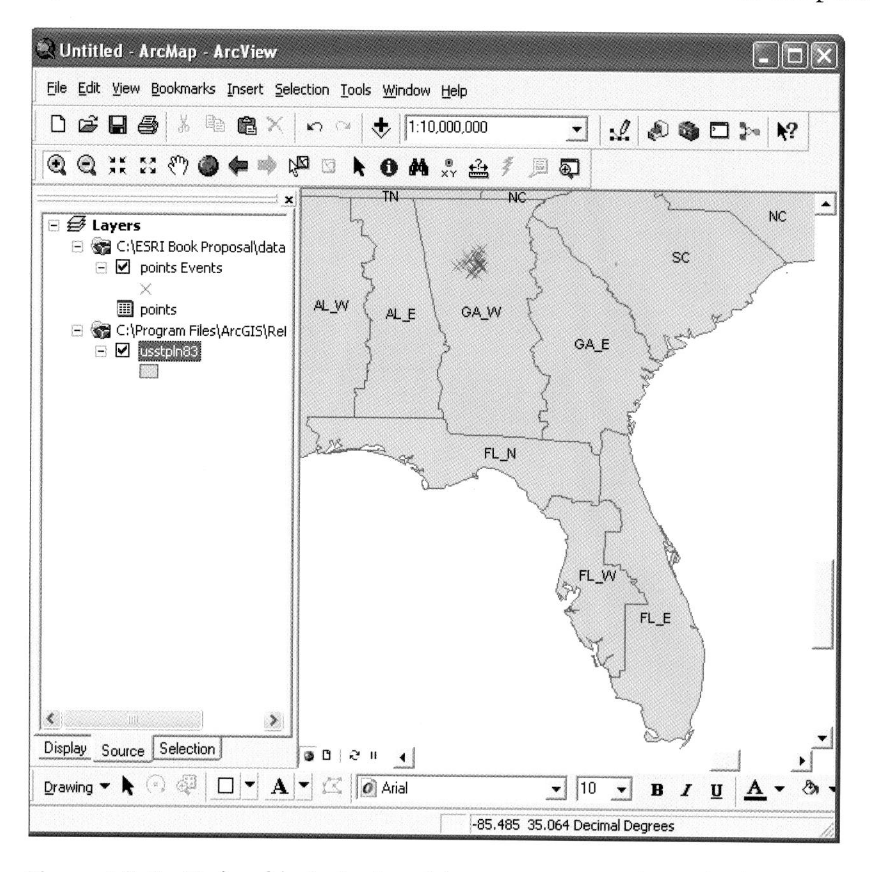

Figure 10-7 Display of the "points Events" theme in ArcMap, overlaying the shapefile usstpln83.

CONVERTING THE X,Y DATA TO SHAPEFILE OR GEODATABASE FEATURE CLASS

The Events theme shown in figure 10-7 has not yet been converted to a shapefile or geodatabase feature class. The data can be maintained as a table, but each time the data is added to ArcMap, the coordinate system would have to be defined over again. It is much more useful to convert the data from the table to a shapefile or geodatabase feature class. Here are the steps.

- Right-click the name of the Events layer, and select Data > Export Data from the drop-down menu, as shown in figure 10-8.

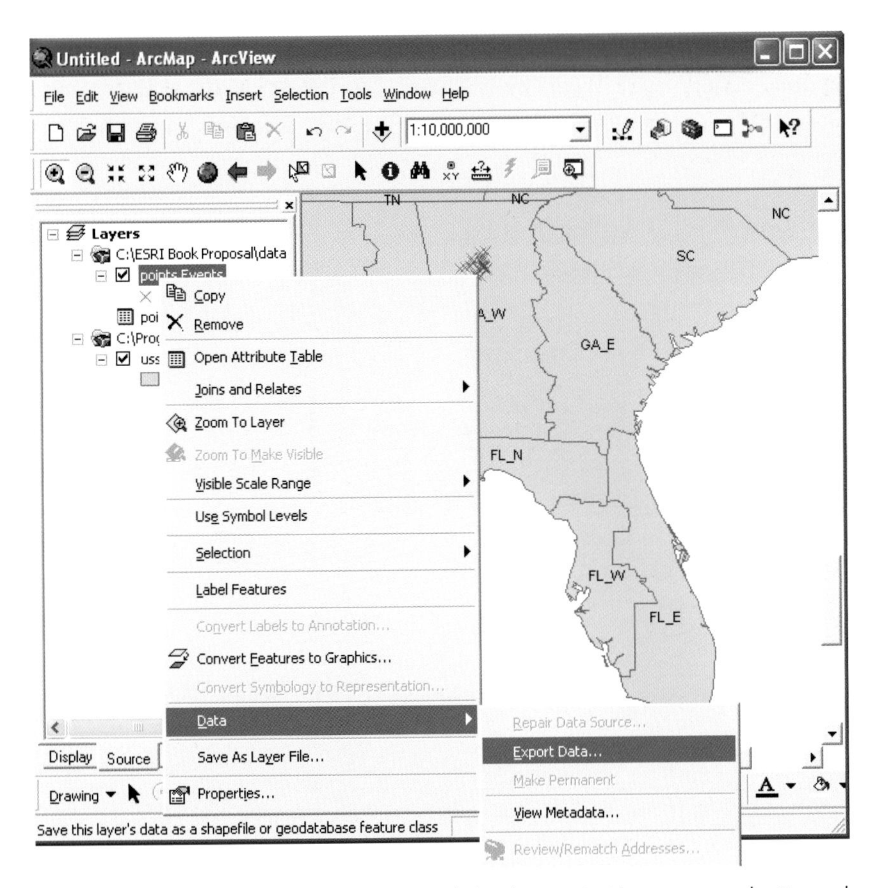

Figure 10–8 Accessing the Export Data dialog box in ArcMap to export the Events theme to a shapefile or geodatabase feature class.

In the Export Data dialog box, select the path to the location where the output data is to be saved, and select the format for the output data. The data can be saved as a shapefile or as a feature class in an existing geodatabase. It is useful to assign a sensible name to the output data instead of the default Export_Output. In this case a name like "georgia_cities_geo83" indicates the kind of data in the output dataset and also provides information about the spatial reference of the data. See figure 10–9 for an example.

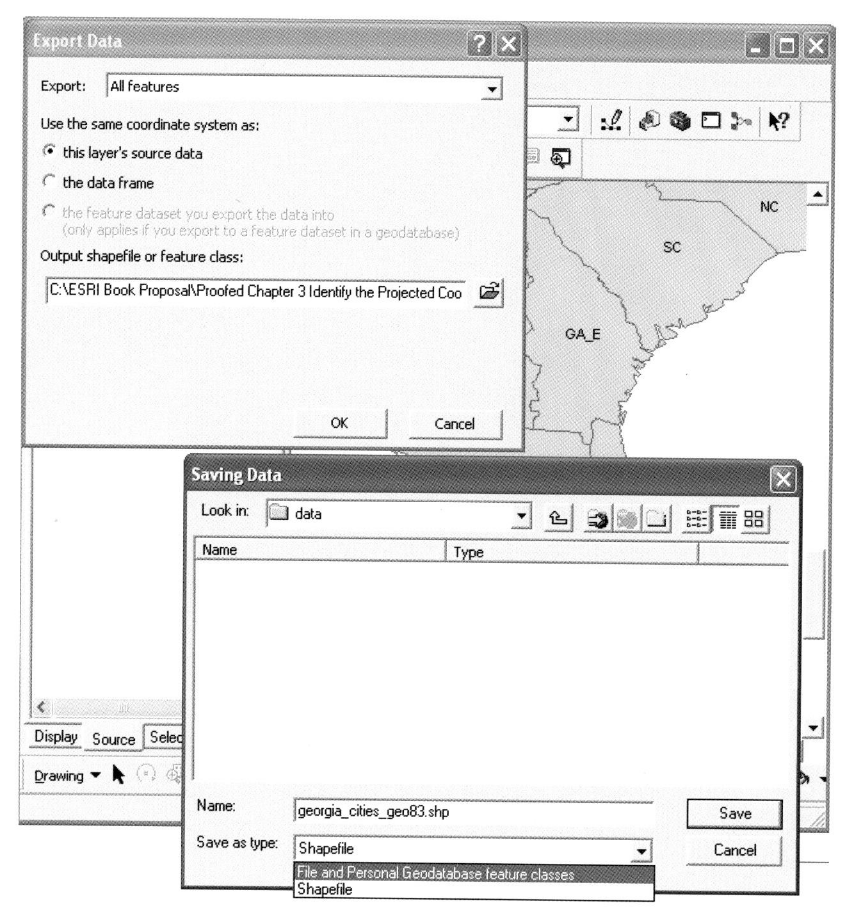

Figure 10–9 In the Export Data and Saving Data dialog boxes, you can browse to the location where the output data will be saved and also select the format—shapefile or geodatabase feature class—for the output dataset.

- Click Save on the Saving Data dialog box, then click OK on the Export Data box. A message will appear asking, "Do you want to add the exported data to the map as a layer?" Click Yes and the new shapefile or geodatabase feature class will be added to the map document.

- You can then remove the Events layer and the original table from the ArcMap data frame by right-clicking those layer names on the Source tab and selecting Remove.

- The new dataset is now available for use in this and other map documents.

WHY BUFFERS DISPLAYED IN ARCMAP ARE NOT ROUND

When creating buffers around point features, then displaying the buffers in ArcMap, the buffers may not appear round or may not actually be circular. The shape of buffers around points in ArcMap is determined by a number of factors:

- The coordinate system assigned to the ArcMap data frame, which may be either geographic or projected.
- The units specified as the buffer radius, which may be linear (feet, meters, etc.) or angular (decimal degrees).
- The coordinate system of the points being buffered.
- The data format of the output buffers.

This section provides a general discussion of factors affecting the buffer display in ArcMap. For further technical discussion, refer to the ESRI Mapping Center blogs named "Buffer Tool" and "Buffer Wizard."

In the following examples, points are being buffered.

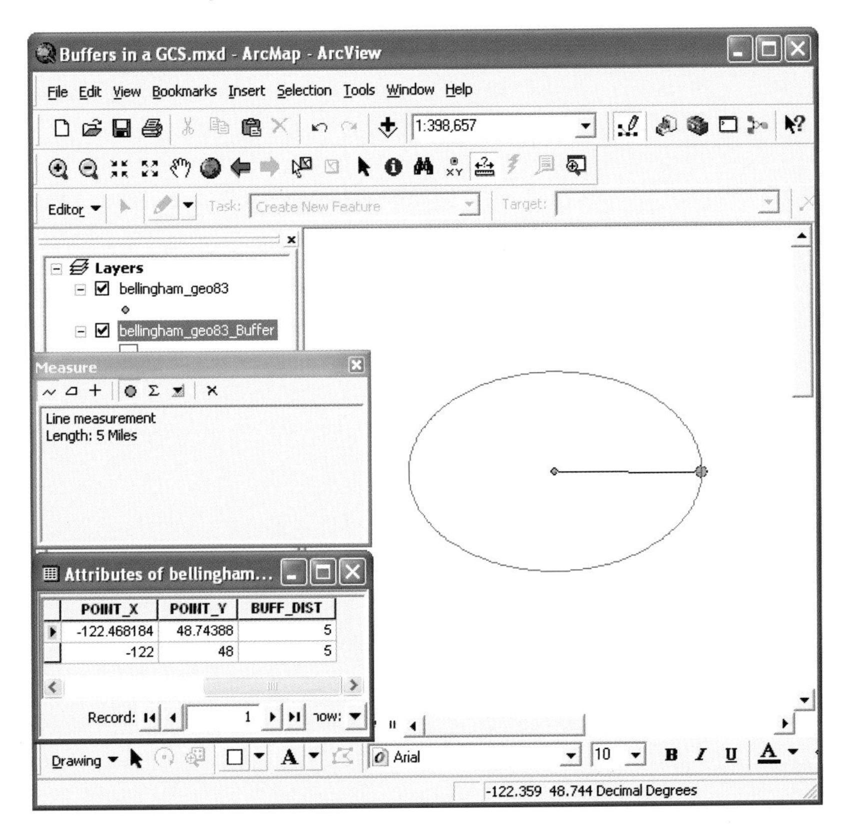

Figure 10–10 The ArcMap display is set to GCS_North_American_1983 and the buffer distance is five miles. Measuring in any direction, the distance from the center point to the buffer is five miles even though the buffer appears oval.

Examining figure 10–10, you see that the point feature from the shapefile that is in geographic coordinates on the NAD 1983 datum and is displayed in the ArcMap data frame in GCS_North_American_1983 has been buffered using a buffer distance of five miles. The buffer does not appear round; the shape of this buffer is decidedly oval. However, using the Measure tool and setting the distance units to miles, you can measure from the buffered point to the buffer, in any direction, and the distance always measures five miles.

Looking back at figure 10-2 (page 147), you recall that data displayed in a geographic coordinate system is progressively more "stretched" in the east-west direction, the farther north or south the data is from the equator. The North and South Poles are points, but a geographic coordinate system displays distances in angular units—degrees—that do not reflect linear surface distances in the east-west direction properly. Since Bellingham, Washington, the location of this point, is far north of the equator, there is substantial east-west stretch of the data. This results in the oval appearance of the buffer.

Now take a look at figure 10-11. This is the same ArcMap document shown in figure 10-10, but the projection of the ArcMap data frame is now set to NAD 1983 UTM zone 10N. Because the data frame is now set to a projected coordinate system, the same buffer has assumed a round shape.

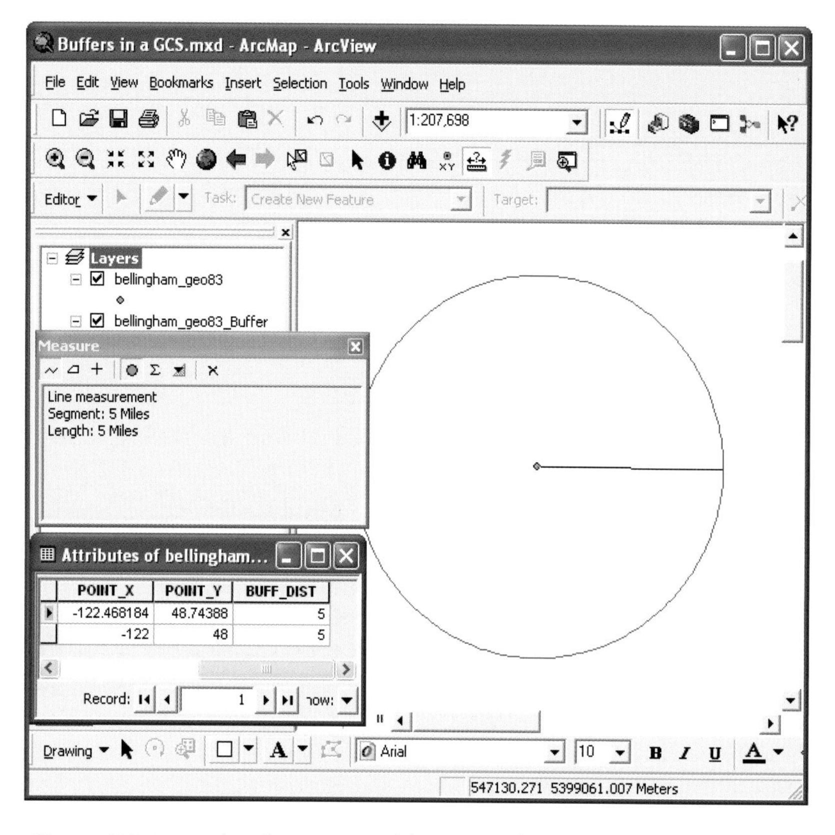

Figure 10-11 When the projection of the ArcMap data frame is changed from the GCS in figure 10-10 to NAD 1983 UTM Zone 10N as shown here, the oval buffer from the previous figure assumes the expected round shape, and the measured distance from the point to the buffer is still five miles.

This time the point shapefile bellingham_geo83.shp is buffered using a angular distance of 0.1 decimal degrees, so the buffer distance units match the units of the projection. Figure 10-12 displays this buffer, with the data frame set to GCS_North_American_1983.

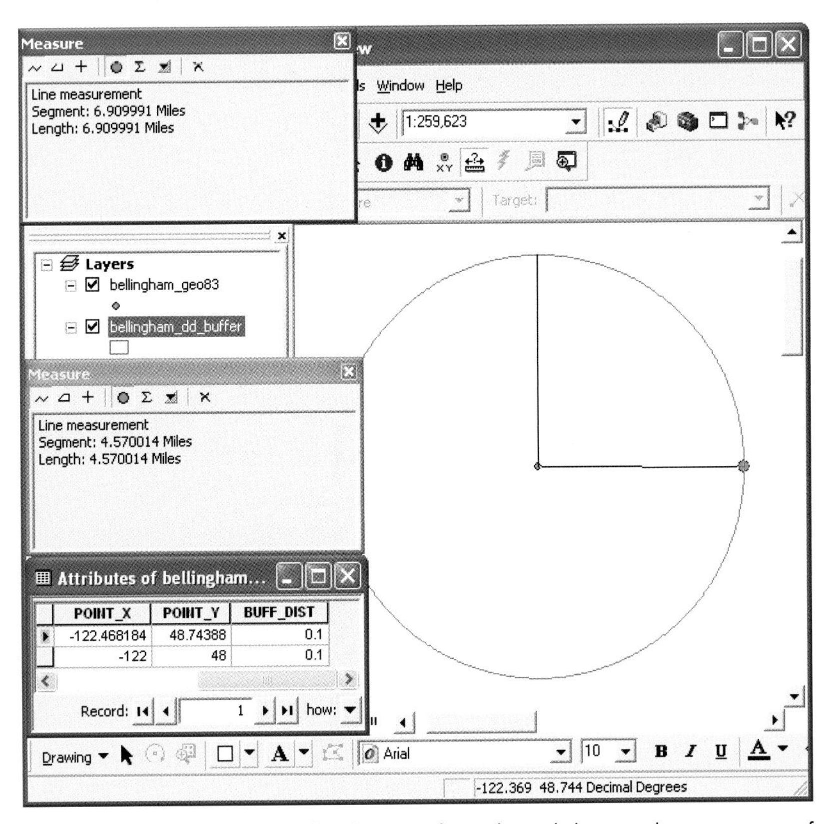

Figure 10–12 Using a buffer distance of 0.1 decimal degrees, the measurement from the center point east to the buffer measures 4.570014 miles. The measurement from the center point north to the buffer measures 6.909991 miles. Even though the display of the buffer is round, the buffer is not geometrically correct as it is displayed in figures 10–10 and 10–11.

In figure 10–13, both the buffer created with linear units of miles and the buffer created with an angular unit of 0.1 degrees are displayed together in a geographic coordinate system. Compare this with figure 10–10, in which the five-mile buffer (red) is also displayed in a GCS. On the next page, the red buffer that is geometrically correct displays as an oval, while the 0.1 degree buffer (blue) displays as a circle, even though we verified that the east-west buffer distance is not the same as the distance in the north-south direction shown in figure 10–12.

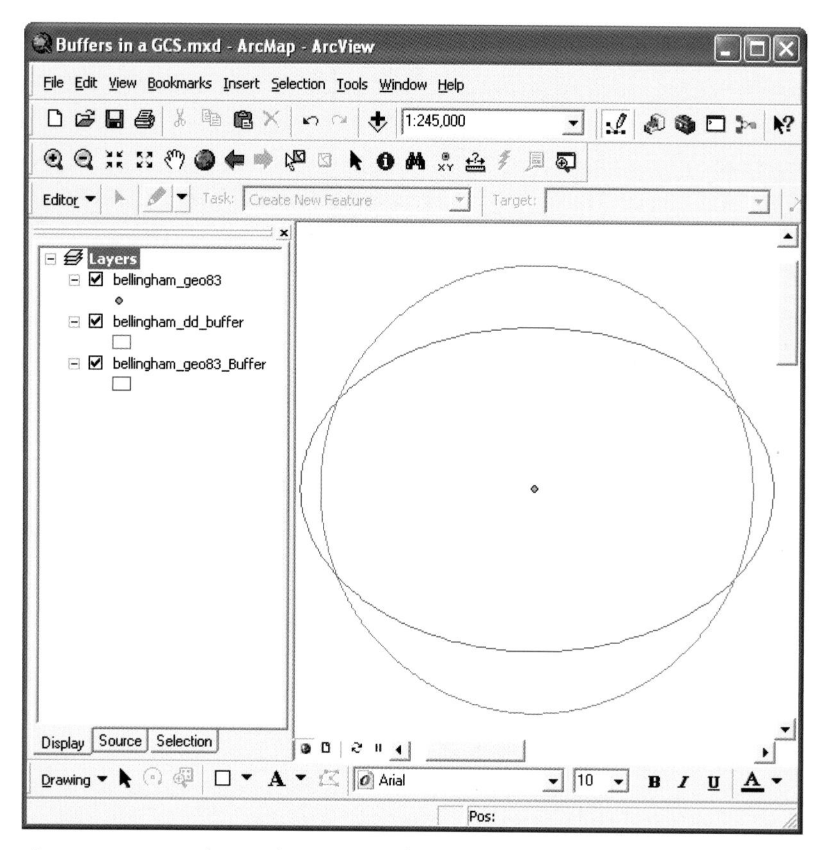

Figure 10-13 The coordinate system of the ArcMap data frame is set to a GCS. The red buffer, created with a linear buffer distance of five miles, appears as an oval even though this buffer is geometrically correct and measures five miles from the center point to the buffer in all directions. The blue buffer, created with the angular distance of 0.1 degrees, appears round even though the distance from the center point to the buffer varies in all directions.

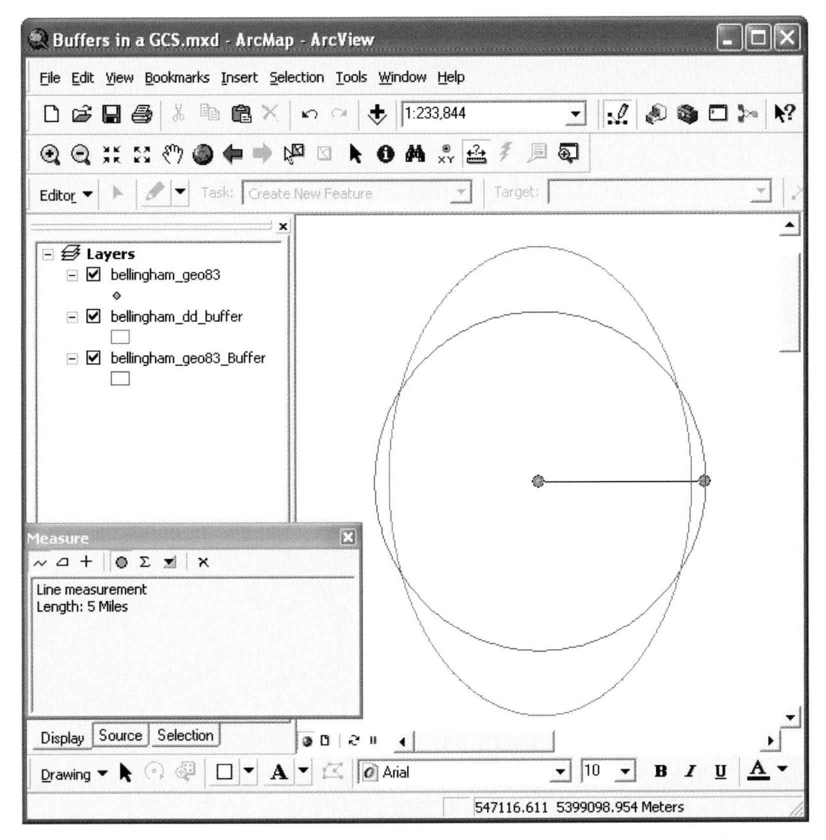

Figure 10–14 Both the five-mile buffer (red) and the 0.1 degree buffer (blue) are displayed with correct proportions when the coordinate system of the ArcMap data frame is set to a projected coordinate system, in this case NAD 1983 UTM Zone 10N. Notice that the blue buffer, which was created with a "distance" in degrees, is compressed east to west. Since we know that lines of longitude converge toward the poles, this is correct behavior.

When buffering data, there are some issues to consider so that the buffers being created will properly serve their purpose.

The coordinate system selected to display the data can cause distortion in buffers. People unfamiliar with that fact often expect buffers to be round. As we have just seen in the figures above, buffers that are geometrically correct can look like ovals if displayed in a GCS. Buffers created with angular units, which are not geometrically correct, appear round when displayed in a GCS.

In order to realize the expectations for round buffers while at the same time creating buffers that are geometrically correct and can be used to analyze data, the best practice is to buffer features with a linear distance unit—feet, miles, meters, kilometers—and to display the buffers in an appropriate projected coordinate system.

In the following illustration, buffers were created around points in a geographic coordinate system, GCS_WGS_1984. The points are 30° apart, east to west as well as north to south. The buffer distance—the radius of the buffers—was specified as 15°.

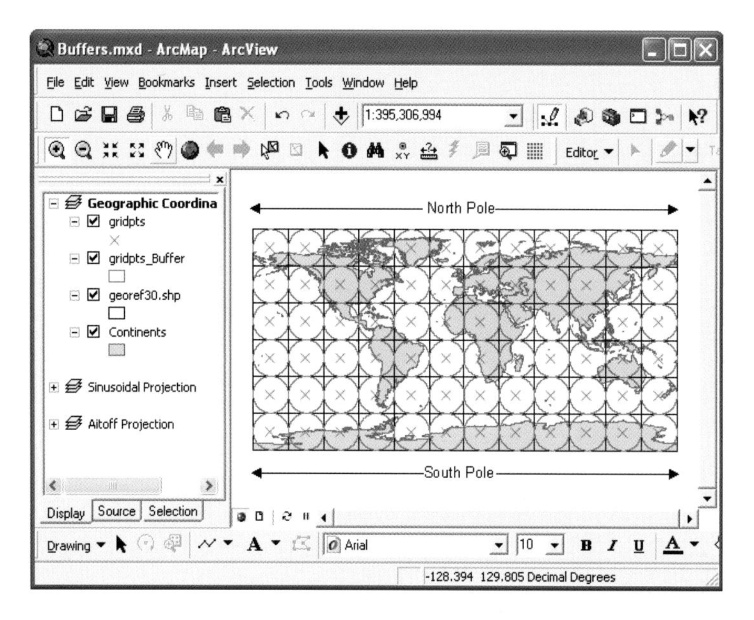

Figure 10–15 A 30° grid covering the entire earth's surface. Each grid "square" is filled with a circle having a radius of 15° and a diameter of 30°.

With the ArcMap data frame set to GCS_WGS_1984 as shown in figure 10–10, the buffers appear round, and all the buffer "circles" appear to be the same size. Recall, though, that when displaying data for the world in a geographic coordinate system, the North and South Poles, which are points, are stretched until both points appear as long as the equator.

The buffers north and south of the equator are *also* stretched in the east-west direction.

To illustrate this, the Measure tool, set to distance units of miles, is used to measure the east-west distance across one 15° grid cell at the equator. The measurement is returned in figure 10–16.

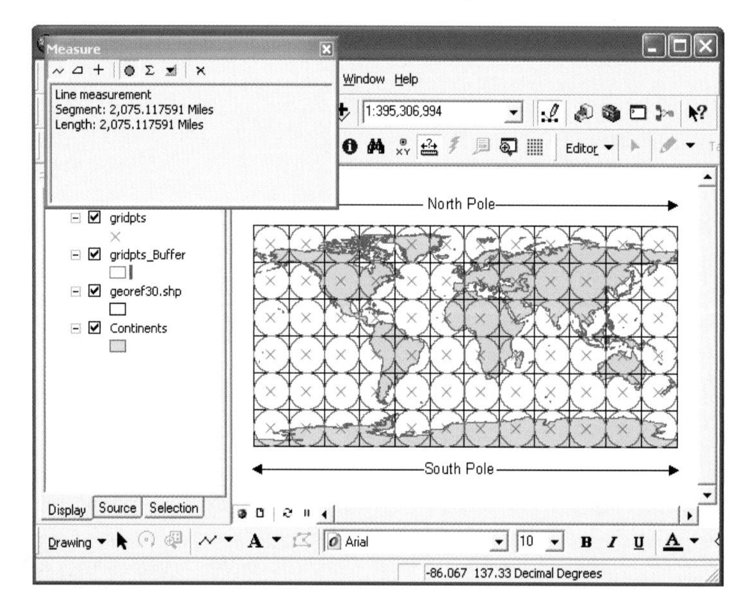

Figure 10–16 Distance measurement of 30° east to west at the equator is about 2,075 miles, as shown in the Measure tool output.

Now, with figure 10-17, let's measure the east-west distance across the buffers close to the North Pole (as shown in figure 10-12).

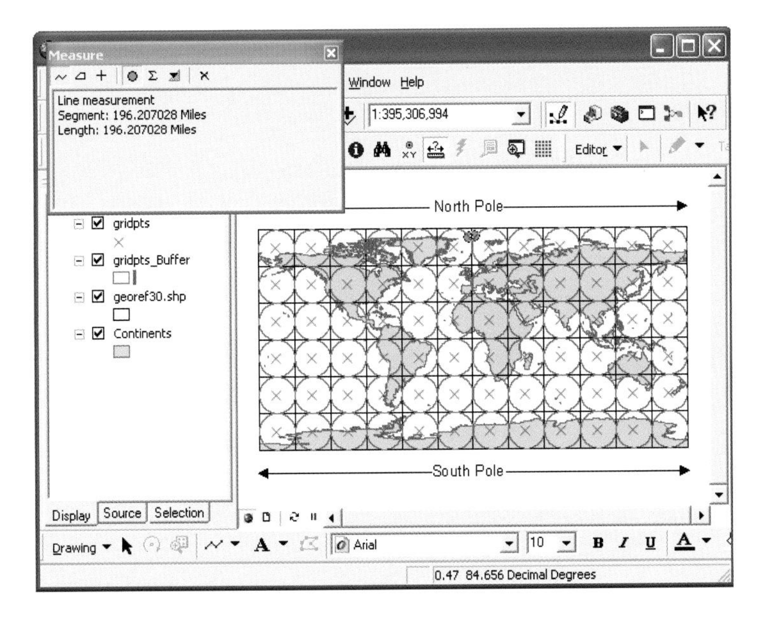

Figure 10-17 The linear distance covered by 30° close to the North Pole is only about 196 miles.

Note the position of the snapping control, and notice that the actual linear distance is now about 196 miles instead of 2,075 miles.

Now change the coordinate system of the ArcMap data frame to a projected coordinate system that illustrates the North and South Poles as points. Applying a sinusoidal-based coordinate system to the ArcMap data frame, and zooming in to a smaller area, you can see the compression of the buffers (as shown in figure 10-13). Instead of displaying as round circles, the buffers now look egg-shaped. The east-west compression of the buffers now illustrates the linear east-west distance across the buffers much more accurately, in figure 10-18.

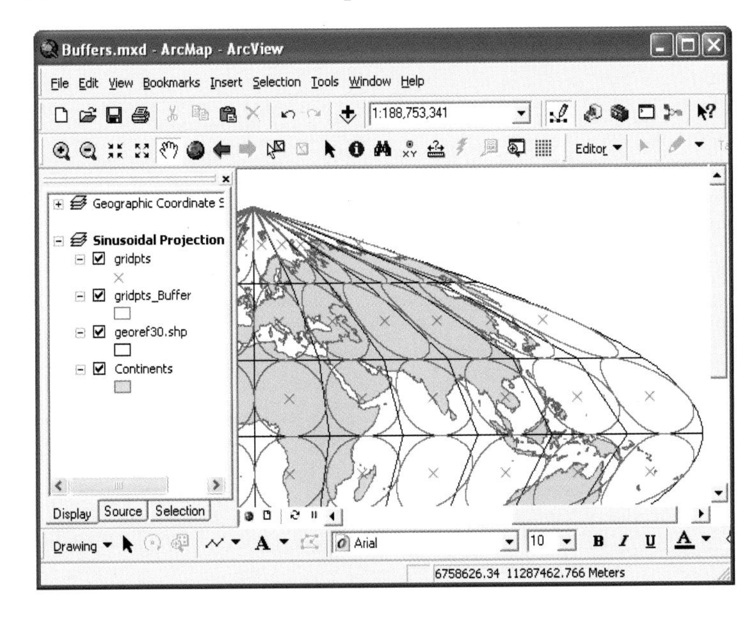

Figure 10-18 Circles with 30° diameters displayed in the sinusoidal projection. Sinusoidal approximates equal area for the entire world. Note that the areas covered by the buffers now show a more realistic representation of each area on the ground.

Aitoff (figure 10–19) is a compromise projection, but can represent areas for the entire world with reasonable accuracy. Again note the change in shape of the buffer circles when zooming in to a quarter of the globe. The North Pole is again represented as a point, although the representation is not as extreme as with the Sinusoidal Projection. (View the latter representation of the data in figure 10–14.)

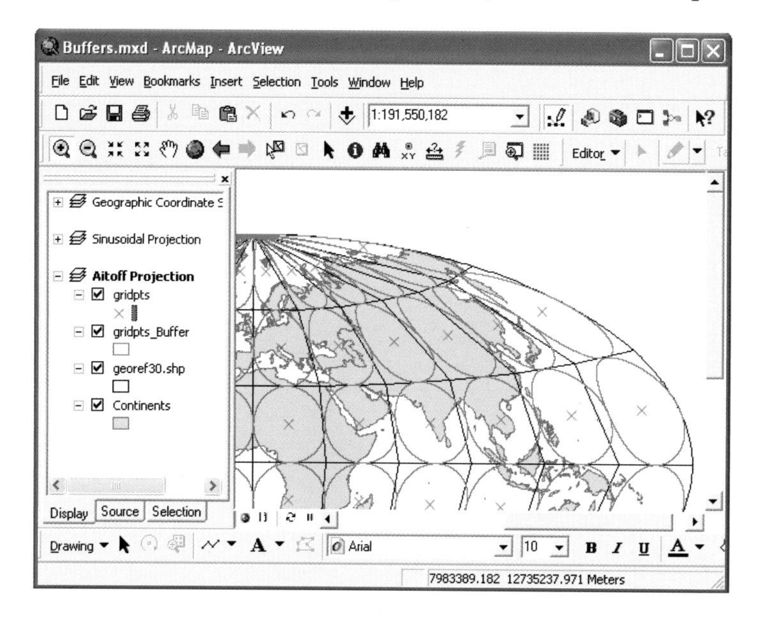

Figure 10–19 Here, the Aitoff projection is used to illustrate the actual surface area of the buffers.

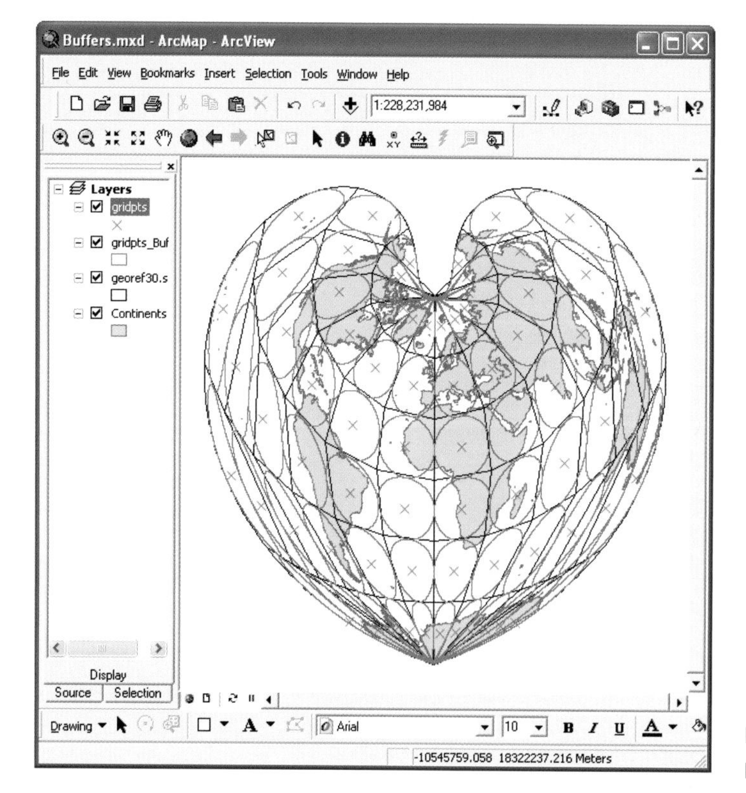

Figure 10–20 Bonne is an equal area projection suitable for use in continental areas.

SUMMARY

This chapter addresses the construction of projection files and explains their parameters in greater detail. The chapter also answers two very common questions:

How do you add x,y data to ArcMap and make the data line up with other data?

Why are buffers created with the tools in ArcGIS Desktop oval instead of round?

Those of us who work in the field of GIS go to our jobs every day, doing our best to make the world a better place. We hope the information presented in this book will be useful to you and will contribute to our shared goal.

APPENDIX A

Knowledge Base articles with critical information referenced in the text

To ensure that you have access to the most current information related to your work with coordinate systems in ArcGIS Desktop, here is a list of Knowledge Base article numbers from the ESRI Support Center. These articles include content or links referred to in the text of this manual.

To access this information, enter **`http://support.esri.com/en/`** in your Web browser.

On the ESRI Support Center home page, in the "Search the Support Center for:" box, enter the article number and click Go. Tables referred to in the text are linked to the relevant articles under the Related Information section at the bottom of each article. Some content is version-specific, so be sure to download the version of the information that applies to your version of ArcGIS Desktop.

While specific references to specific content may occur in the text of the book, it is strongly advisable for you to download and print *all* this information. Having printed copies of these tables that are readily accessible makes working with coordinate systems much easier, eliminates guesswork, and ensures greater accuracy in your mapping and analysis projects.

Article Number	Title
18317	*FAQ: What National Transformation Version 2 geographic transformation grids are supported in ArcGIS Desktop?*
21327	*How To: Select the correct geographic (datum) transformation when projecting between datums* Link title: ArcGIS Projection Engine <version> Datum transformation methods and appropriate geographic areas
22455	*How To: Convert Degrees Minutes Seconds values to Decimal Degree values using the Field Calculator*
24646	*How To: Select a suitable map projection or coordinate system* Link title: Projections table
27548	*How To: Convert a file with coordinates in Degrees, Minutes, and Seconds to a shapefile using ArcMap (9.3.1 and prior versions only)*
29280	*FAQ: What geographic coordinate system or datum should be used for my data?* Link title: Geographic Coordinate Systems and Area of Use <version>
30727	*How To: Format a table in Microsoft Excel for use in ArcMap*
35152	*How To: Install a new NTv2 geographic transformation and grid file for use in ArcGIS Desktop*
37264	*How To: Convert a file with coordinates in Degrees—Minutes—Seconds to a shapefile using ArcMap 10*

APPENDIX B

Default Install Paths for ArcGIS Desktop versions 9; 9.1; 9.2; 9.3/9.3.1

Supported Operating System	Default Installation Path
Windows 2000 Professional	*C:\Program Files\ArcGIS*
Windows 2003 (32-bit) Server Standard, Enterprise, Datacenter	*C:\Program Files\ArcGIS*
Windows 2003 (64-bit (EM64T)) Server Standard, Enterprise, Datacenter	*C:\Program Files\ArcGIS*
Windows 2008 (32-bit) Server Standard, Enterprise, Datacenter	*C:\Program Files\ArcGIS*
Windows 2008 (64-bit (EM64T)) Server Standard, Enterprise, Datacenter	*C:\Program Files\ArcGIS*
Windows 7 (32-bit) Ultimate, Professional, Home Premium	*C:\Program Files\ArcGIS*
Windows 7 (64-bit (EM64T)) Ultimate, Professional, Home Premium	*C:\Program Files\ArcGIS*
Windows Vista (32-bit) Ultimate, Enterprise, Business, Home Premium	*C:\Program Files\ArcGIS*
Windows Vista (64-bit (EM64T)) Ultimate, Enterprise, Business, Home Premium	*C:\Program Files\ArcGIS*
Windows XP (32-bit) Professional Edition, Home Edition	*C:\Program Files\ArcGIS*
Windows XP (64-bit (EM64T)) Professional Edition, Home Edition	*C:\Program Files\ArcGIS*

Default Install Paths for ArcGIS Desktop version 10

Supported Operating System	Default Installation Path
Windows 2003 (32-bit) Server Standard, Enterprise, Datacenter	*C:\Program Files\ArcGIS\Desktop10.0*
Windows 2003 (64-bit (EM64T)) Server Standard, Enterprise, Datacenter	*C:\Program Files\ArcGIS\Desktop10.0*
Windows 2003 Server Terminal Services	*C:\Program Files\ArcGIS\Desktop10.0*
Windows 2008 (32-bit) Server Standard, Enterprise, Datacenter	*C:\Program Files\ArcGIS\Desktop10.0*
Windows 2008 (64-bit (EM64T)) Server Standard, Enterprise, Datacenter	*C:\Program Files\ArcGIS\Desktop10.0*
Windows 7 (32-bit) Ultimate, Professional, Home Premium	*C:\Program Files\ArcGIS\Desktop10.0*
Windows 2008 (64-bit (EM64T)) Server Standard, Enterprise, Datacenter	*C:\Program Files\ArcGIS\Desktop10.0*
Windows 7 (32-bit) Ultimate, Professional, Home Premium	*C:\Program Files\ArcGIS\Desktop10.0*
Windows 7 (64-bit (EM64T)) Ultimate, Professional, Home Premium	*C:\Program Files\ArcGIS\Desktop10.0*
Windows Vista (32-bit) Ultimate, Enterprise, Business, Home Premium	*C:\Program Files\ArcGIS\Desktop10.0*
Windows Vista (64-bit (EM64T)) Ultimate, Enterprise, Business, Home Premium	*C:\Program Files\ArcGIS\Desktop10.0*
Windows XP (32-bit) Professional Edition, Home Edition	*C:\Program Files\ArcGIS\Desktop10.0*
Windows XP (64-bit (EM64T)) Professional Edition, Home Edition	*C:\Program Files\ArcGIS\Desktop10.0*

APPENDIX C

Default User Profile Path to Coordinate Systems folder for ArcGIS Desktop 9.3/9.3.1

Supported Operating System	Default User Profile Path to Coordinate Systems Directory
Windows 2000 Professional	*C:\Documents and Settings\<user_name>\Application Data \ESRI\ArcMap\Coordinate Systems*
Windows 2003 (32-bit) Server Standard, Enterprise, Datacenter	*C:\Documents and Settings\<user_name>\Application Data \ESRI\ArcMap\Coordinate Systems*
Windows 2003 (64-bit (EM64T)) Server Standard, Enterprise, Datacenter	*C:\Documents and Settings\<user_name>\Application Data \ESRI\ArcMap\Coordinate Systems*
Windows 2008 (32-bit) Server Standard, Enterprise, Datacenter	*C:\Users\<user profile>\AppData\Roaming\ESRI\ArcMap \Coordinate Systems*
Windows 2008 (64-bit (EM64T)) Server Standard, Enterprise, Datacenter	*C:\Users\<user profile>\AppData\Roaming\ESRI\ArcMap \Coordinate Systems*
Windows 7 (32-bit) Ultimate, Professional, Home Premium	*C:\Users\<user profile>\AppData\Roaming\ESRI\ArcMap \Coordinate Systems*
Windows 7 (64-bit (EM64T)) Ultimate, Professional, Home Premium	*C:\Users\<user profile>\AppData\Roaming\ESRI\ArcMap \Coordinate Systems*
Windows Vista (32-bit) Ultimate, Enterprise, Business, Home Premium	*C:\Users\<user profile>\AppData\Roaming\ESRI\ArcMap \Coordinate Systems*
Windows Vista (64-bit (EM64T)) Ultimate, Enterprise, Business, Home Premium	*C:\Users\<user profile>\AppData\Roaming\ESRI\ArcMap \Coordinate Systems*
Windows XP (32-bit) Professional Edition, Home Edition	*C:\Documents and Settings\<user_name>\Application Data \ESRI\ArcMap\Coordinate Systems*
Windows XP (64-bit (EM64T)) Professional Edition, Home Edition	*C:\Documents and Settings\<user_name>\Application Data \ESRI\ArcMap\Coordinate Systems*

Default User Profile Paths to Coordinate Systems folder for ArcGIS Desktop version 10

Supported Operating System	Default User Profile Path for Coordinate Systems
Windows 2003 (32-bit) Server Standard, Enterprise, Datacenter	*C:\Documents and Settings\<user_name>\Application Data \ESRI\ArcMap\Coordinate Systems*
Windows 2003 (64-bit (EM64T)) Server Standard, Enterprise, Datacenter	*C:\Documents and Settings\<user_name>\Application Data \ESRI\ArcMap\Coordinate Systems*
Windows 2008 (32-bit) Server Standard, Enterprise, Datacenter	*C:\Users\<user profile>\AppData\Roaming\ESRI\ArcMap \Coordinate Systems*
Windows 2008 (64-bit (EM64T)) Server Standard, Enterprise, Datacenter	*C:\Users\<user profile>\AppData\Roaming\ESRI\ArcMap \Coordinate Systems*
Windows 7 (32-bit) Ultimate, Professional, Home Premium	*C:\Users\<user profile>\AppData\Roaming\ESRI\Desktop10.0 \ArcMap\Coordinate Systems*
Windows 7 (64-bit (EM64T)) Ultimate, Professional, Home Premium	*C:\Users\<user profile>\AppData\Roaming\ESRI\Desktop10.0 \ArcMap\Coordinate Systems*
Windows Vista (32-bit) Ultimate, Enterprise, Business, Home Premium	*C:\Users\<user profile>\AppData\Roaming\ESRI\Desktop10.0 \ArcMap\Coordinate Systems*
Windows Vista (64-bit (EM64T)) Ultimate, Enterprise, Business, Home Premium	*C:\Users\<user profile>\AppData\Roaming\ESRI\Desktop10.0 \ArcMap\Coordinate Systems*
Windows XP (32-bit) Professional Edition, Home Edition	*C:\Documents and Settings\<user_name>\Application Data \ESRI\Desktop10.0\ArcMap\Coordinate Systems*
Windows XP (64-bit (EM64T)) Professional Edition, Home Edition	*C:\Documents and Settings\<user_name>\Application Data \ESRI\Desktop10.0\ArcMap\Coordinate Systems*

FURTHER READING

Flacke, Werner, and Birgit Kraus. *Working with Projections and Datum Transformations in ArcGIS, Theory and Practical Examples.* 2005. Points Verlag Norden Halmstad, Germany.

Snyder, John P. *Map Projections Used by the U.S. Geological Survey.* 1982, 2nd ed. 1983. Geological Survey Bulletin 1532, USGS, Washington, D.C.

Wade, Tasha, and Shelly Sommer. 2006. *A to Z GIS, An Illustrated Dictionary of Geographic Information Systems.* ESRI Press, Redlands, California.

DATA SOURCE CREDITS

Chapter 1 screen shots include:
geographiccoordinates.jpg, derived from World Data Bank II, courtesy of ArcWorld

Chapter 2 screen shots include:
Ccbuildings_noprojection.shp, courtesy of City of Riverside, California. Information Technology Department
Cczoning_geo.shp, courtesy of City of Riverside, California. Information Technology Department

Chapter 3 screen shots include:
usstpln27.shp, from ESRI Data & Maps 2008, courtesy of NOAA, USGS, ESRI
usstpln83.shp, from ESRI Data & Maps 2008, courtesy of NOAA, USGS, ESRI
utm.shp, derived from World Data Bank II, courtesy of ArcWorld
continents.shp, derived from World Data Bank II, courtesy of ArcWorld

Chapter 4 screen shots include:
Ccbuildings_noprojection.shp, courtesy of City of Riverside, California. Information Technology Department
Cczoning_utm83.shp, courtesy of City of Riverside, California. Information Technology Department
Parcels.dwg, courtesy of GIS Department/Rockingham Planning
Index.dwg, courtesy of GIS Department/Rockingham Planning

Chapter 5 screen shots include:
N/A

Chapter 6 screen shots include:
Parcels.dwg, courtesy of GIS Department/Rockingham Planning
Index.dwg, courtesy of GIS Department/Rockingham Planning

Chapter 7 screen shots include:
cntry2006.sdc, from ESRI Data & Maps 2006, courtesy of ArcWorld Supplement
nadcon_grid_extents.shp, digitized by the author
NTv2_grid_sample.shp, digitized by the author
NTv2_grid_sample2.shp, digitized by the author
NTv2_grid_sample3.shp, digitized by the author

Chapter 8 screen shots include:
Neighborhoods_sp83, courtesy of City of Riverside, California. Information Technology Department
CanyonCrest_spHARN, courtesy of City of Riverside, California. Information Technology Department
Neighborhoods_geo83, courtesy of City of Riverside, California. Information Technology Department
CanyonCrest_geo27, courtesy of City of Riverside, California. Information Technology Department

Chapter 9 screen shots include:
continents.shp, derived from World Data Bank II, courtesy of ArcWorld
georef15.shp, derived from World Data Bank II, courtesy of ArcWorld
longlat.bmp, derived from World Data Bank II, courtesy of ArcWorld
utm.shp, derived from World Data Bank II, courtesy of ArcWorld

Chapter 10 screen shots include:
continents.shp, derived from World Data Bank II, courtesy of ArcWorld
georef10.shp, created by the author
points.dbf, created by the author
usstpln83.shp, from ESRI Data & Maps 2004, courtesy of NOAA, USGS, ESRI
bellingham_geo83.shp, created by the author
bellingham_geo83_Buffer.shp, created by the author
bellingham_dd_buffer.shp, created by the author
gridpts.shp, created by the author
georef30.shp, created by the author
gridpts_Buffer.shp, created by the author

INDEX

D

E

F

G

H

I

K

L

M

N

O

P

Q

R

S

T

U

V

W

X

Y

Z

Related titles from ESRI Press

GIS Tutorial 2: Spatial Analysis Workbook

ISBN: 978-1-58948-258-6

Updated for ArcGIS 10, *GIS Tutorial 2* offers hands-on exercises to help GIS users at the intermediate level continue to build problem-solving and analysis skills. Inspired by the *ESRI Guide to GIS Analysis* book series, this book provides a system for GIS users to develop proficiency in various spatial analysis methods, including location analysis; change over time, location, and value comparisons; geographic distribution; pattern analysis; and cluster identification.

The ESRI Guide to GIS Analysis, Volume 1: Geographic Patterns and Relationships

ISBN: 978-1-87910-206-4

The ESRI Guide to GIS Analysis, Volume 1, demonstrates how geographic analysis with GIS can identify patterns, relationships, and trends that lead to better decision making. This book focuses on six of the most common geographic analysis tasks: mapping where things are, mapping the most and least, mapping density, finding what is inside, finding what is nearby, and mapping what has changed.

Also available: *The ESRI Guide to GIS Analysis, Volume 2: Spatial Measurements and Statistics*

ISBN: 978-1-58948-116-9

Designing Geodatabases: Case Studies in GIS Data Modeling

ISBN: 978-1-58948-021-6

This highly visual guide to creating a dynamic geographic data model helps ArcGIS users design schemas that have comprehensive and descriptive query definitions, user-friendly cartographic displays, and increased performance standards. This work outlines five steps for taking a data model through its conceptual, logical, and physical phases—modeling the user's view, defining objects and relationships, selecting geographic representations, matching geodatabase elements, and organizing the geodatabase structure.

Unlocking the Census with GIS

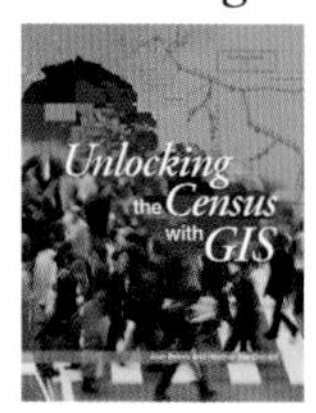

ISBN: 978-1-58948-113-8

Unlocking the Census with GIS describes how GIS can be used to better access, understand, manage, and analyze census data and census-related information and present it in a spatial format. Numerous maps, tables, sidebars, and other in-depth examples and explanations are provided to guide readers to an understanding of the census and its value to those using powerful GIS software tools.

ESRI Press publishes books about the science, application, and technology of GIS. Ask for these titles at your local bookstore or order by calling 1-800-447-9778. You can also read book descriptions, read reviews, and shop online at www.esri.com/esripress. Outside the United States, visit our Web site at www.esri.com/esripressorders for a full list of book distributors and their territories.